Lecture Notes in Computer Science   3457

Commenced Publication in 1973
Founding and Former Series Editors:
Gerhard Goos, Juris Hartmanis, and Jan van Leeuwen

T0224638

Michael Smirnov (Ed.)

Autonomic Communication

First International IFIP Workshop, WAC 2004
Berlin, Germany, October 18-19, 2004
Revised Selected Papers

 Springer

Volume Editor

Michael Smirnov
Fraunhofer FOKUS
Kaiserin-Augusta-Allee 31, 10589 Berlin, Germany
E-mail: smirnow@fokus.fraunhofer.de

Library of Congress Control Number: 2005928376

CR Subject Classification (1998): C.2, H.3, H.4

ISSN   0302-9743
ISBN-10   3-540-27417-0 Springer Berlin Heidelberg New York
ISBN-13   978-3-540-27417-9 Springer Berlin Heidelberg New York

Springer is a part of Springer Science+Business Media

springeronline.com

© 2005 IFIP International Federation for Information Processing, Hofstrasse 3, 2361 Laxenburg, Austria
Printed in Germany

Typesetting: Camera-ready by author, data conversion by Scientific Publishing Services, Chennai, India
Printed on acid-free paper   SPIN: 11520184   06/3142   5 4 3 2 1 0

Preface

The first IFIP Workshop on Autonomic Communication (WAC 2004) was held 18–19 October 2004 in Berlin, Germany. The workshop was organized by Fraunhofer FOKUS with the help of partners of the EU-funded Autonomic Communication Coordination Action — IST-6475 (ACCA), and under technical sponsorship of IFIP WG6.6 — Management of Networks and Distributed Systems.

The purpose of this workshop was to discuss Autonomic Communication—a new communication paradigm to assist the design of the next-generation networks. WAC 2004 was explicitly focused on the principles that help to achieve purposeful behavior on top of self-organization (self-management, self-healing, self-awareness, etc.). The workshop intended to derive these common principles from submissions that study network element's autonomic behavior exposed by innovative (cross-layer optimized, context-aware, and securely programmable) protocol stack (or its middleware emulations) in its interaction with numerous, often dynamic network groups and communities. The goals were to understand how autonomic behaviors are learned, influenced or changed, and how, in turn, these affect other elements, groups and the network. The highly interactive and exploratory nature of WAC 2004 defined its format — six main sessions grouped in three blocks, each block followed by a panel with all speakers of the previous block as panellists and session chairs as panel moderators.

The first panel aimed to highlight the main principles guiding research in algorithms, protocols and middleware; the second panel investigated grand challenges of network and service composition; the third panel had to answer the question "How Does the Autonomic Network Interact with the Knowledge Plane?". Panel reports were compiled by panel moderators and conclude this volume.

The emphasis of the workshop was on the long-term research agenda with a broad interdisciplinary approach to explore concurrently multiple paradigm spaces. Along with regular sessions the workshop offered two invited talks. Prof. I. Chlamtac, IEEE and ACM Fellow, known for outstanding achievements in optical and mobile networking, shared his views on bio-inspired communication. Prof. A. Fradkov, IEEE Fellow and IFAC member, known for his fundamental works on non-linear and adaptive control, presented a novel network control paradigm based on a cyberphysical approach. Both talks are published in this volume in full length in the invited program section.

The motivating vision of autonomic communication is that of a self-organized Internet that will be able to sense its environment, to perceive its changes and to understand the meaning of these changes, thus facilitating new ways to perform network control, management, middle box communication, service creation, etc. This might be based on universal and fine-grained multiplexing of numerous policies, rules and events which, while autonomously performed, can facilitate the desired behavior of groups of network elements. In line with this vision papers were solicited on topics including policy-based communication and policy multiplexing, group communication for the control and management plane, network evolvability design, self-organization for re-configurability, management of nomadicity, autonomic communication calculi and theoretical foundations of autonomic network control, mobile code and network programmability, generic network-level service composition at run-time, context handling, theoretical foundations of rule-based systems, security, immunity and resilience of autonomic communication, and its application to QoS, traffic engineering, routing, etc. The workshop received 45 submissions from all over the world; the TPC selected 18 papers constituting the main body of this volume.

WAC 2004 was the foundational event for the autonomic communication initiative and this volume is believed to be the first collective publication solely dedicated to the investigation into its principles. The initiative emerged from the series of brainstorming and consultation meetings started at the EU Commission premises in July 2003 and organized by the IST Programme Future and Emerging Technologies (FET) to address Communication Paradigms for 2020. WAC 2004 was followed by the foundation meeting of the ACF — Autonomic Communication Forum (http://www.autonomic-communication-forum.org), and by the publication of FP6 Call for Projects within the proactive initiative on Situated and Autonomic Communication. At the time of writing ACF had close to 200 members from industry and academia. WAC 2005 is being organized in Athens, Greece; WAC 2006 is planned for Paris, France.

To give the authors the opportunity to revise their papers based on the workshop discussions, to allow panel chairs to discuss and to prepare panel reports, and to allow invited speakers to publish their presentations as full papers, this LNCS volume was published as a postproceedings of the workshop.

Finally, it is a pleasure to record here our high appreciation of the efforts of many people in the successful launch of WAC: to all the authors who submitted, presented and revised their papers, or agreed to present their papers as posters, regretting that it was not possible to accept more papers for WAC 2004; to all the attendees for highly interactive participation in the discussions; to the Program Committee members and to all associated reviewers for thorough and motivated assessment of submissions; to partners in the Autonomic Communication Coordination Action for acting as promoters and helpers for WAC 2004; to the EU Commission FET officers for the continuous support of WAC and ACF; to IFIP TC6 members for accepting this foundational workshop, especially to IFIP WG6.6 for technical hosting of WAC; and last but not least, many warm thanks go to the employees of Fraunhofer FOKUS, who dedicated much effort in making the event professionally organized and socially enjoyable.

April 2005   Michael Smirnov

About This Book

This is the postproceedings of the 1st IFIP TC6 WG6.6 International Workshop on Autonomic Communication (WAC 2004); it includes 18 full papers presented at WAC 2004 and revised by the authors based on the workshop discussions, and full texts of the two invited talks and three panel reports.

Workshop Chairs

General Chair   Popescu-Zeletin, R. (Germany)
TPC Chair   Smirnov, M. (Germany)

Program Committee

Biersack, E., Eurecom, France
Boutaba, R., Waterloo U., Canada
Bindel, F., T-Com, Germany
Campbell, A., Columbia U., USA
Chapin, L., ACM, USA
Diot, C., Intel, UK
Dobson, S., UCD, Ireland
Denazis, S., Hitachi, France
Einsiedler, H., DT AG, Germany
Fall, K., Berkeley U., USA
Farserothu, J., CSEM, Switzerland
Fdida, S., UPMC, France
Fradkov, A., IPME, Russia
Haring, G., U. Vienna, Austria
Hutchison, D., Lancaster U., UK
Karlsson, G., KTH, Sweden
Koufopavlou, O., U. Patras, Greece
Leduc, G., ULG, Belgium
Liebeherr, J., U. Virginia, USA
Matta, I., Boston U., USA

Morris, R., MIT, USA
Mulvenna, M., U. Ulster, UK
Nixon, P., UCD, Ireland
Plattner, B., ETHZ, Switzerland
Pujolle, G., UPMC, France
Seneviratne, A., UNSW, Australia
Schieferdecker, I., TUB, Germany
Sestini, F., EU Commission, EU
Smirnov, M., FOKUS, Germany
Spaniol, O., Aachen U., Germany
Stavrakakis, I., NKUA, Greece
Sterrit, R., U. Ulster, UK
Tafazolli, R., Surrey U., UK
Tschudin, C., Basel U., Switzerland
Vicente, J., Intel, USA
Wolf, L., Braunschweig U., Germany
Yamamoto, L., Hitachi, France
Zitterbart, M., Karlsruhe U.,
    Germany

Local Organizing Committee (Fraunhofer FOKUS)

Benner, B.

Vladova, G.

Wollgramm, I.

Lohmar, K.

Baharlou, B.

Jaenke, C.

Burkhardt, F.

Schulzke, F.

Quandel, G.

Lange, B.

Scheddin, D.

Winkler, J.

Radke, C.

Cuno, S.

Kum, W.

Huang, Y.

Sponsoring Institutions

Fraunhofer FOKUS

EU Commission, FET Programme

IFIP WG6.6

Table of Contents

Negotiation and Deployment

Immunity and Resilience

Meaning, Context and Situated Behaviour

Invited Programme

Panel Reports

An Infrastructure-Based Approach to Support Dynamic Networks with Mobile Agents

Arndt Döhler, Christian Erfurth, and Wilhelm Rossak

Computer Science Dept., Friedrich-Schiller-University,
Jena, 07743 Jena, Germany
{arndt.doehler, erfurth, rossak}@informatik.uni-jena.de
http://swt.informatik.uni-jena.de/

Abstract. With the growing size of distributed systems and the higher number of available resources and services in networks dynamical aspects become more and more important in systems engineering. We believe that there is a real need for decentral, self-organizing structures to cope with the upcoming challenges. In this paper we describe a framework which provides a self-organizing infrastructure that allows to link otherwise autonomous elements in a flexible way and adapts dynamically to changes in the underlying network. This framework is implemented as an extension of the mobile agent system Tracy, which is also a product of our university. The Tracy Domain Management module is part of the framework and provides the basis for segmenting the infrastructure. Another module we are going to discuss in this paper facilitates autonomous and proactive routing of mobile agents. Agents form the application layer of the system. Routing is triggered by the needs an agent inherits from its owner and then matched to the resources and services available in the network in an iterative fashion. We describe concepts, design issues and first results of our work with Tracy and the use of these additional Tracy modules.

Keywords: Distributed systems, self-organization, rule-based behavior, proactive navigation of mobile agents, mobile agent systems.

1 Introduction

In the context of networked environments, mobile agents can be seen as a new paradigm for the implementation of fully distributed software systems with a balanced peer-to-peer concept [1]. During the last years at Friedrich-Schiller-University Jena (FSU), we have developed our own mobile agent system (MAS) Tracy [2, 3]. Tracy is a Java2-based middle-ware that supports the efficient migration of mobile agents over several protocols and migration strategies. So called agencies (Tracy agent servers) are the specialized execution environments for mobile agents. In our approach, every Java-enabled device in the Internet can be such a network node. Currently, we work on additional system components on top of the basic middle-ware layer to network mobile agencies by a self-organizing

M. Smirnov (Ed.): WAC 2004, LNCS 3457, pp. 1–12, 2005.
© IFIP International Federation for Information Processing 2005

mechanism, to improve scalability and flexibility, and to provide an information base for mobile agents that supports their pro-activity and adaptability. Especially interesting is the case where the network provides a dynamical environment [4], e. g. if mobile network nodes and services appear and disappear, and where agents act as intelligent entities by determining their own path at run-time dynamically in the continuously changing landscape.

The movement of mobile agents is based on a logical network view, i. e. mobile agents discern agencies only. The cooperation of normally autonomous and independent agencies is essential to network agencies on such a logical level. The first part of this paper covers that issue and describes a self-organizing network of agencies.

The second part of the paper addresses the routing service which improves the movement of mobile agents in such networks and supports their autonomy. On an agent's journey, it visits only those agencies which provide a resource or service of interest. Furthermore, the agent tries to use a fast path through a network based on known infrastructure characteristics (as QoS). Finally, an agent optimizes its transmissions between agencies with the help of several migration strategies described by Braun [5]. All information necessary for the agent's navigation in the network and the related calculations are provided by the routing service module.

2   Concepts of the Basic Infrastructure

2.1   A Logical Network

A node with an agency is the basic element of our infrastructure. All networked agencies form a logical or virtual application-level network. Every agency offers services managed by local stationary agents. Mobile user-task agents (application-agents) can use these services by local message exchange with the stationary agents. To use remote services on other agencies, a mobile agent must migrate to the desired agencies for local communication with the stationary agents. This approach is typical for a strictly defined MAS and has been described e. g. in Braun [5].

In this context an autonomous decision of a mobile agent is based on a couple of basic capabilities each agency must exhibit: Knowledge regarding the existence of other agencies and theirs offered services is essential, the propagation of this information through the network is desirable, and the infrastructure must be enabled to handle network changes.

The problem is, that in the worst case every agency would have to hold information regarding every other known agency and, thus, a fully intermeshed virtual network comes into being. Since fully intermeshed networks aren't a scalable solution in industrial size networks, we decided to separate the network into manageable and interrelated chunks.

2.2   Topology – The Domain Concept

The basic idea of our approach is to split the whole MAS network into domains (see Fig.1), which are limited to IP-subnetworks. All agencies within a domain register at a single agency called domain manager. In our approach all agencies have basically fully equal rights and basic capabilities since the DomainInformationAgent, the domain management component of Tracy, is present on each agency. So we have a peer-to-peer system. By launching an agency as a domain manager it takes on a specific role and offers the relevant domain management services. From the network management view this role-based behavior can be seen as a client-server behavior, where the domain manager plays the role of the server.

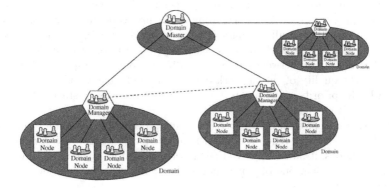

Fig. 1. Domain Concept: A structured network of agencies

The domain manager is responsible to manage all other agencies in a domain called domain nodes and to hold connections to other domains. Every domain node has its unique domain manager, and the domain manager knows all domain nodes that are currently active in its domain. If an agency starts or stops, it has to register respectively check out with the domain manager node.

The domain manager holds the complete information of its domain nodes and of itself. It propagates this information to all domain nodes inside its domain, but not beyond. Thus, all agencies within the domain know each other, and form again a fully intermeshed network of a limited size. In practical applications we have learned to expect not more than 60 agencies per domain.

To re-integrate the whole logical agent system network, domains have to be linked together. For that reason domain managers contact a unique domain master. The master is a specialized domain manager which manages only other domain managers and interconnects them with each other. In future, we plan some more domain masters to prevent the single point of failure problem. On this level, only summarized information are exchanged.

Since it is possible to launch more than one domain, this approach is capable of handling very large networks in a piece by piece fashion, while it allows for scalability at the same time. Only inside a domain information and resources are fully intermeshed. In-between domains the mesh is broken. This makes it, of course, necessary for a mobile agent to move into a domain before it can access its resources.

2.3   Valency of a Node – Priority Concept

A Tracy domain is a self-organized basic cell of the infrastructure. When a Tracy agency is launched, it checks the presence of other agencies by sending a UDP-multicast first. If a domain manager answers, the agency must register with the domain manager by sending a mobile agent. In the case of absence of a domain manager the agency becomes domain manager itself. If several agencies were launched simultaneously or a domain manager breaks down, agencies compete to become the domain manager according to the first-come-first-serve principle.

To influence the role allocation according to importance of an agency, a priority value can be assigned to every agency [6]. The priority is modeled as a byte value and ranges between -128 and $+127$. It should correspond to the performance, the quality of the network connection, and the reliability of a node. Currently the priority value has to be fixed before the agency starts. After the launch it can't be changed.

With the concept of priorities, the launching process of a domain information agent changes slightly. When a domain manager receives registration messages from other nodes, it now compares their priorities with its own value. If its own priority is lower than one of a new node, that node becomes the new domain manager.

The drawback of this solution is the fixed assignment of priority values. Furthermore, the programmer has to know the absolute valency of his device or the valency in ratio to the other agencies before its agency starts. This leads to an arbitrary or appraised allocation of the priority value.

2.4   Valency of a Node – Dynamical Priority Assignment

Our new approach is, therefore, to dynamically assign proper priority values during the runtime of an agency. Launching an agency happens as described before, but after registration with the domain manager respectively become domain manager itself an agency performs performance measurements (computing power, memory size and others) by some sensors. The performance measurements reflect the performance of the node and form an abstract view on the local capabilities of the system environment of the agency.

The Map Module which will be discussed in section 3.1 provides information on known services and on network connection qualities. Together with the performance measurement results as an information base (see Fig. 2) it is now possible to calculate a proper priority value to support an automated and useful choice of a domain's manager.

Performance measurements can be regularly repeated and the time interval can be dynamically adapted to the current network situation with the help of small statistics on the last measured values. If there are changes in the network accessibility (e. g. by a more badly signal-noise-ratio of a 802.11 connection) or the usage rate of a node (e. g. by a higher utilization by other applications) an agency will take notice of it and the priority can be changed.

If the transfer rate decreases or the RTT to a node increases, or a node's computing power decreases and the memory utilization increases, the node is less suitable to manage a domain because this leads to additional utilization in computing power, memory and network load. So the priority of such a node has to decrease (and vice versa) according to the relative changes of the values.

If a service is launched on a node, it must be checked if it is an administrative service or an application service. In the first case the priority has to be decreased because of direct, additional agency utilization. In the second case the service may be performed by an application outside of the agency. From the performance perspective this application does not directly increase the load the agency has to handle. Therefore, the additional utilization of the host platform should be measured by the sensors. Thus, the start of an application service means mostly that the host platform is a stable and reliable computer with an excellent network connection. This more logical hint can't be measured or performed otherwise but ought to be observed over time. If the reliability assumption is affirmed, priority can be increased.

2.5   Dynamical Priority Assignment – Scenario

A typical scenario is shown in Fig. 2 and describes the usefulness of dynamical priority assignment. On the left side the priorities of the domain nodes are predefined, static and without any relation to the logical network's real situation. From the bottom upwards there are three layers which corresponds to three logical views. A network quality view comes from the Map Module which is fed by several network sensors. A view of the logical agency network and the roles of the known agencies comes from the DomainInformationAgent itself. Several node performance and utilization sensors feed the DomainInformationAgent directly. Note, that the sensors are not shown in the Figure. The most abstract logical view forms the top layer of this Figure: the service view.

After calculating the priority each node sends the value to the domain manager. The most prioritized node becomes the new domain manager, shown on the right side of the Figure.

2.6   Discussion

The dynamical priority assignment represents an abstract closed control loop. The priorities are the control variables and the role allocation of the domain manager is the controlled system. An inherent design problem of closed control loops is to prevent unstable and instable states, which may caused by too strong feedbacks of the controlled system or by disturbances. In our case an instable state means a recurrent shift of the domain manager role. By the role changing itself, the priority of the new domain manager will decrease due to the utilization

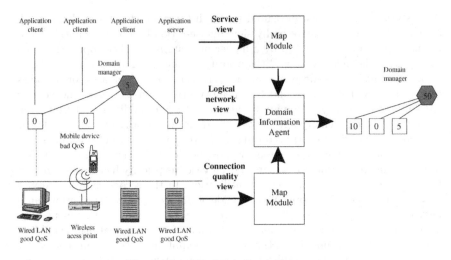

Fig. 2. Dynamical Priority Assignment

by its new domain management service. To prevent unstable states, we stabilize the system by introducing a programmable threshold value that must be reached by a new domain manager candidate. Currently we program a fixed threshold value of 20%. Additionally we use small statistics on older priority values to get a time-dependent change of the priority which stabilize the control loop as well. A drawback of this solution is the relatively slow change of the domain manager role in situations with high network dynamics.

3   Proactive Navigation

In modern computer networks services can be regarded as dynamical components. To be able to use services, a mobile agent is in need of information about service location and reachability. To answer this need, we have developed a framework called *ProNav*. Its most important feature is to locate services and information in the network and to offer this type of data to any mobile agent currently planning its itinerary. This is achieved by integrating the data that is locally acquired by each agent server into a so-called m ap that enables each agent to recognize and analyze its virtual environment. Even more, an agent is able to adapt to environmental changes without human intervention. These mechanisms utilize the domain concept, as discussed above, as a basic feature and extend it with additional functionality.

From an architectural perspective, *ProNav* extends any MAS by working as an intermediate layer in-between the actual agent system and the application layer that is formed by specialized mobile agents and their user and application interfaces (see Fig. 3).

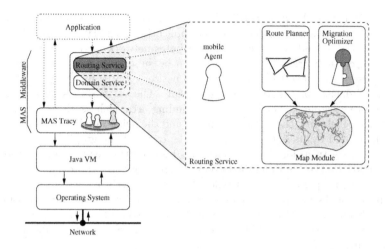

Fig. 3. Architectural overview of *ProNav* and integration as middle-ware

In Figure 3, an architectural overview of the additionally introduced system components is presented. These components are integrated into the MAS Tracy using stationary agents. In general, such agents are not able to migrate but offer local services. Mobile agents are able to use local services by employing agent to agent communication within the local agency.

ProNav is divided into three major components: the M ap M odule, the R oute P lanner, and the M igration O ptim izer. In principle each component may be used independently by any mobile agent. However, only by integrating their services a mobile agent will achieve full autonomy and pro-activity for the itinerary planning task.

The Map Module is used by a mobile agent to locate services and to access information on network connection qualities. Connection qualities are especially important for the Route Planner and the Migration Optimizer to achieve optimization. The Route Planner calculates a "short" path through the network. The Migration Optimizer optimizes each single migration included in an agent's itinerary from a more technological, in our case Tracy-specific, efficiency perspective. This module is mainly designed to reduce network load by selecting and transmitting only those code and data portions of the agent that are needed at the upcoming remote agencies. This is, if necessary, done by a concept called slicing [7]. Other options are to place code in advance in the network, to send data home to carry less "luggage", to change the transmission protocol, etc. The Optimizer is not focus of this paper [8].

3.1 A Logical Network Map

Building on the Domain Service, *ProNav* collects information to generate a "network map" offering information to mobile agents. To achieve this, we implemented a Map Module which consists of several network sensors and a map data structure. In addition to information on application-level services provided by the agencies, this module collects and distributes network status information.

The logical network of agencies needs to be subdivided using the Domain Service described above to achieve scalable network maps. Basically, a map of an agency consists of a partial network graph. The vertices of such a partial graph are the visible agencies of the surrounding area such as all nodes in the local domain including the domain manager and the neighbored remote domains each represented by its domain manager. The edges of the graph represent the end-to-end view transport layer connections between the vertices. Each edge is characterized by the "full qualified domain name" of the remote agency and a couple of network parameters that reflect the current performance of the end-to-end connection. The Map Module uses network sensors with interfering measurement methods on top of Java to get the characteristics of a connection. There are sensors to measure availability, latency, transfer rate, and transmission time of a standard agent.

As an example, we describe the function of the latency sensor which measures round trip times of a minimal data packet. This means the sensor emulates a PING over a TCP connection. The sensor opens a connection to a special port of a remote agency. On this special port the remote sensor listens for measurement requests. After establishing the connection, the sensor starts the time measurement and sends a small packet. The answer of the remote agency is an acknowledgment, the measurement stops and the connection is canceled. After a definable duration a new measurement with the next agency will start.

We have made a set of evaluation measurements to get a feeling of the sensor's quality. In Fig. 4 the measured values of the latency sensor are compared with values delivered by the PING of the OS in a wireless environment (IEEE 802.11b WLAN) and an Ethernet environment (IEEE 802.3 10BASE-T 10 mbit/s half duplex). The values of the sensor correspond to PING. Due to the application level implementation of sensors the values are a little bit above the PING values.

To flatten peaks measured by sensors, we use forecast modules which generate next expected values on basis of a small time series of measured values. The value of the forecast module which has delivered the best forecast in the last run is entered into the map.

The transfer rate sensor works in almost the same manner. However, for a measurement larger data packets are needed. To transmit useful data thereby these packets are used to exchange and propagate gathered map data (service offers, QoS) between agencies. As a result, every agency has complete information about connection qualities and service offers within its domain. A Domain Map is created. The domain manager summarizes its Domain Map information and propagates this compressed information to other known domains. So within a domain, every agency has a network map with detailed information about the local domain and relevant information about known remote domains (via the domain master). Thus a mobile agent is able to locate services within a remote domain. To utilize such a service, the agent has to migrate to the domain manager

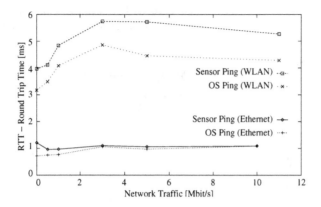

Fig. 4. Round trip times: Latency sensor vs. OS-PING

of the remote domain, access the local Domain Map, which is different from the current one, and finally migrate to the actual service location.

3.2   Route Planning

The Route Planner organizes an agent's trip through the network of agencies. The route planning process itself is basically the Traveling Salesman Problem (TSP) [9] which is a NP-complete type of problem. As a consequence, getting an optimal solution in practical application is ruled out. But there are heuristic algorithms (such as local search, genetic, simulated annealing, neural network algorithms etc.) that have been applied extensively for solving such problems [10]. The comparative performance of the algorithms depends on the problem and the given detailed circumstances.

The calculation of an itinerary is based on the map data. We calculate a kind of distance matrix simply by using the reciprocal values of measured transfer rate. This matrix has to be updated at regular time intervals to fit the environment's dynamical behavior. Then, a path finder algorithm is applied in order to get a distance matrix with shortest paths between places (without short cuts). In some experiments, we figured out that our distance matrix is not symmetrically in general. This is caused by oscillating transfer rates values and non symmetrical connections like DSL. For TSP, there are algorithms for asymmetrical (ATSP) [11] and for symmetrical matrices (STSP) [12].

In our case, local optimization algorithms are a good choice. Hence, our route planning process starts with a nearest neighbor search algorithm to generate an initial path through the net. This path is input for further optimizations with an adapted version of the iterated 3-Opt algorithm (I3Opt). In Figure 5, the result of the nearest neighbor algorithm is about 36% above optimum (optimum means minimum in this case) but is calculated within 0.7 ms (Pentium II 333 with Java). This route planning is done on a generated matrix of the problem space tm at (triangulated random matrices) with 100 places [13]. Such a matrix is an asymmetrical one where an entry is the shortest path between two places.

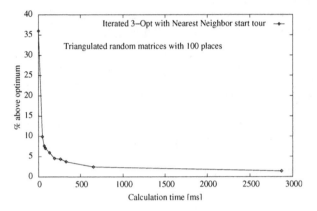

Fig. 5. Route planning with tmat100

To avoid unnecessary calculation, we compare the so far calculated migration time with the path improvement during the last steps. If the time benefit of the last 20 ms calculation is not greater than the path improvement the calculation stops. Thereby it takes also the calculation power into account.

The result of the calculation is an agent's initial itinerary. During an agent's journey it might be useful, or even necessary, to modify this itinerary (changed network status, new services, etc.). This can be done by the agent itself without any human-agent interaction.

3.3   A Sample Scenario

The following scenario describes the application of *ProNav* in a network of agencies. Thereby a mobile agent visits a set of agencies while migrating through the network to fulfill its task.

A user (the owner) hands over a task to an agent. Normally, such a task should not contain information on H O W to fulfill. Hence, the agent has to organize the journey through the network by itself. Therefore, the agent searches for suitable services in the map provided by the local agency. This map contains information on services within the domain and some network characteristics. The search result is a set of agencies that should be visited. Now the agent may trigger the Route Planner to use the available map's information on connection topology and qualities to identify a possible trip through the network. The result is a first travel plan – the itinerary. Before the agent begins the trip, it might use the Migration Optimizer to optimize the trip from an efficiency perspective. Now the agent "executes the itinerary" and starts the migration. During the trip the agent visits service points and communicates and cooperates with other agents.

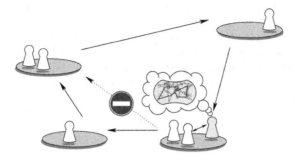

Fig. 6. Proactive navigation of a mobile agent in a dynamical environment

At any point in time, but at least when migrating to further away agencies (the map's information is more blurred for further away agencies), the agent may fine-tune and re-adapt its itinerary. This is achieved by taking advantage of information now available in different domains. Finally, after its trip, the agent hands over the results to its principal. This might include a description of the visited agencies, a kind of travel report.

As indicated, we want to provide an infrastructure that enables agents to be more autonomous. A user should concentrate on W HAT the agent has to do and not on HOW .

4   Conclusion

We see mobile agent systems as one of the more promising alternatives to develop truly distributed systems in large and dynamic networking environments. For a mobile agent the opportunity for a proactive and autonomous planning of its itinerary is essential in this context. This feature is offered by *ProNav* as part of a generic infrastructure framework. The agent's programmer and user do no longer have to plan the itinerary for the agent. The agent is enabled to fulfill this task itself and independently of its owners. *ProNav* also provides enough information and flexibility to abandon the notion of a fixed route through the network and allows for regular updates and changes in the itinerary during its execution. This helps to react immediately and in an autonomous fashion to changes in the environment.

A basic robust infrastructure organization is important for *ProNav* to function as described. However, logical or virtual networks exhibit a high level of internal dynamics: Agencies and services are added, deleted, or modified, connection quality changes over time, etc. Therefore, the Domain Concept was introduced and enhanced with new priority functionalities to better react to the dynamics of the system and to provide a basis for the *ProNav* module.

The introduced approaches, the Domain Concept and *ProNav*, are in general not limited to MASs. They can be used to enable the self-organization of any autonomous distributed system that supports a minimum of communication and autonomy.

As a next step we plan to model the dynamic behavior of the system formally. We also plan to use control loop approaches, well known in electrical engineering, to analyze the effects of dynamical priority assignment. We also look at other infrastructures for distributed systems to go beyond the current MAS-based implementation.

References

1. Vigna, G.: Mobile Code Technologies, Paradigms, and Applications. PhD thesis, Politecnico di Milano (1998)
2. Friedrich-Schiller-Universität Jena, Software Engineering Group: Tracy – The Mobile Agent System. URL: http://tracy.informatik.uni-jena.de (2004)
3. Braun, P., Erfurth, C., Rossak, W.: An Introduction to the Tracy Mobile Agent System. Technical Report Math/Inf/00/24, Friedrich-Schiller-Universität Jena, Institut für Informatik (2000)
4. Erfurth, C., Rossak, W.: Characterization and Management of Dynamical Behaviour in a System With Mobile Agents. In Unger, H., Böhme, T., Mikler, A., eds.: Innovative Internet Computing System - Second International Workshop, IICS 2002, Kühlungsborn (Germany), June 2002. Volume 2346 of Lecture Notes in Computer Science., Kühlungsborn (Germany), Springer Verlag (2002) 109–119
5. Braun, P.: The Migration Process of Mobile Agents - Implementation, Classification, and Optimization. PhD thesis, Friedrich-Schiller-Universität Jena, Institut für Informatik (2003)
6. Braun, P., Eismann, J., Rossak, W.: A Multi-Agent Approach To Manage a Network of Mobile Agent Servers. Technical Report 12/01, Friedrich-Schiller-Universität Jena, Institut für Informatik (2001)
7. Fensch, C.: Class Splitting as a Method to Reduce Network Traffic in a Mobile Agent System. Master's thesis, Friedrich-Schiller-Universität Jena, Institut für Informatik (2001)
8. Schaaf, M.: Entwicklung und Implementierung einer Komponente zur Migrationsoptimierung für Mobile Agenten. Master's thesis, Friedrich-Schiller-Universität Jena, Institut für Informatik (2003)
9. Lin, S.: Computer Solutions of the Traveling Salesman Problem. Bell System Technical Journal **44** (1965) 2245–2269
10. Johnson, D.S., McGeoch, L.A.: The Traveling Salesman Problem: A Case Study in Local Optimization. In E.H.L.Aarts, J.K.Lenstra, eds.: Local Search in Combinatorical Optimization. John Wiley and Sons, Ltd. (1997) 215–310
11. Johnson, D.S., Gutin, G., McGeoch, L.A., Yeo, A., Zhang, W., Zverovitch, A.: Experimental analysis of heuristics for the atsp. [14] 445–487
12. Johnson, D.S., McGeoch, L.A.: Experimental analysis of heuristics for the stsp. [14] 369–444
13. Cirasella, J., Johnson, D.S., McGeoch, L.A., Zhang, W.: The Asymmetric Traveling Salesman Problem: Algorithms, Instance Generators, and Tests. In: Proceedings of ALENEX. (2001)
14. Gutin, G., Punnen, A.P., eds.: The Traveling Salesman Problem and its Variations. Kluwer Academic Publishers (2002)

Some Requirements for Autonomic Routing in Self-organizing Networks

Franck Legendre, Marcelo Dias de Amorim, and Serge Fdida

LIP6/CNRS – Université Pierre et Marie Curie,
8, rue du Capitaine Scott – 75015 – Paris – France
{legendre, amorim, sf}@rp.lip6.fr

Abstract. This paper addresses some requirements of self-organizing networks as well as interoperability problems due to merges and splits phenomena. In a mobile environment, merges and splits characterize the spatial overlap between two self-organized networks. While merge refers to the time when two disjoint networks meet and overlap, split refers to the time of partition. In a dynamic environment, AutoComm (AC) principles bring a new support for interoperability since current protocol heterogeneity is observed at all stack layers from the radio interface to applications. In this paper, we reconsider the formalization of a community and its requirements. We then characterize the split and merge phenomena and their implications. We give some requirements that must fulfill solutions to merging (high context-awareness) in order for AC groups to self-scale. Finally, we propose a merging solution for overlapping wireless self-organized networks using heterogeneous routing protocols.

1 Introduction

Current networks are limited by principles edicted 30 years ago when requirements of mobility did not exist. Since that time, several innovations were proposed to bypass inherent limitations of IP principles (e.g., NAT, Mobile IP, IPv6). Moreover due to the dynamic nature of ad hoc networks, principles driven by end-to-end requirements do no longer apply. New innovative routing paradigms must be designed.

In fact, routing is the basic service of a network and any other service resides on this fundamental functionality. Hence, we believe routing requires being adaptable to the diverse environments, new usages, and QoS requirements desired by applications. Two opposite directions respond to this new constraint. The first approach claims that different routing schemes suit different contexts and that one routing protocol fits all cannot be envisioned. Currently, this approach has led to an heterogeneous set of routing protocols. The second approach proposes new flexible communication paradigms such as i3 [1] and Network Pointers [2], which are adapted to new constraints brought by mobility. With i3 the act of sending is decoupled from the act of receiving. Addresses are based on a communication identifier either unicast or multicast. Network

M. Smirnov (Ed.): WAC 2004, LNCS 3457, pp. 13–24, 2005.

pointers extend the semantic of addresses from basic identifiers to specific packet handling functions that are not just limited to forwarding. Both approaches bring innovations but we believe interoperability is required until a unique protocol or paradigm prevails if one can be designed. Indeed, diversity is the right approach to fit the multiplicity of contexts. It is all the more true since the design of routing protocols also relies heavily on the underlying layers and radio access technologies. These technologies are only beginning to develop with Ultra Wide Band (UWB) [3], Multiple Input Multiple output (MIMO) [4], and beam forming with smart antennas. The diversity of solutions (at the routing layer) hence enables to better suit evolution of usages, requirements of the environment and underlying radio technologies. We consider that diversity will be an invariant of forthcoming routing protocols and that AutoComm (AC) will enable to handle efficiently this situation. Diversity managed by AC principles will enable new protocols and paradigms to fusion or be dropped. This process will be similar to living elements' natural selection that uses sexual reproduction to enhance the overall fitness of their genetic patrimony [5].

For example, first developments in wireless routing were the adaptation of wired protocols (e.g., RIP, OSPF) giving birth to OLSR [6] and reactive protocols (e.g., DSR [7] and AODV [8]) which are direct adaptations of the Address Resolution Protocol (ARP) to multi-hop wireless networks. However, combining these two approaches lead to more efficient and adaptive protocols such as ZRP [9] and SHARP [10]. AC's framework brings a new chance to reinvest routing. We must restart the cycle of defining routing protocols dedicated to specific purposes and conditions. This might lead possibly to combined solutions and enhance the overall benefit.

Nethertheless, this evolution shows limitations of current proposals for ad hoc routing. These relate to two correlated factors. First, the model of the wireless channel was considered similar to the wired Internet and researchers only considered new requirements brought by mobility. Up to now, evolutions in networking where direct application of wired technologies to the wireless world without rethinking the basics. People saw wireless ad hoc networks as wired networks with end-to-end requirements and hence narrowed their vision required to bypass such a limited model. Second, such limitations stem from the fact that the network architecture as it is designed nowadays reflects our incapacity of communicating between people involved in lower layers with people involved in upper layers. IP, by its universal goal of unifier marks the barrier between both sides and is the point of convergence. This leverages the question of what is now the advantage of such a rigid model in a changing environment. Cross-layering brings a first response to bypass limitations of the layering paradigm. AC is to come next.

AC proposes to reinvest a research effort to bypass these limitations by proposing a dynamic framework and an interdisciplinary view of all networking aspects including routing. AC is a new opportunity to avoid rehearsing the same mistakes. We believe, however, that the design of AC-compliant network elements will not respect in the short term AC's design rules and philosophy. We believe that a first step will combine existing solutions while introducing AC components. The final step will be the design of communicating elements fully AC-compliant. New networking functionalities must be designed accordingly to AC principles but still interoperate with current existing protocols. For example, with the current diversity of ad hoc and wired routing protocols, interoperability has not been much tackled. We think that the goal of AC is limited, among other purposes to enable protocol interoperation. In our case, we focus on routing protocol interoperation. This temporary solution will at term leave place to a single yet completely dynamic framework for routing that will support the emergence of innovative paradigms.

In this paper, we address the issue of one of the main challenging problems in next generation networks, namely routing in merge and split environments. We will mainly study the merge phenomenon and its implications, and define requirements of how an AC must react to merges. Routing has mainly been designed to cope with the scaling relative to the number of individuals collaborating in a group and thus cope with expanding networks (e.g., Internet growth, wireless network radio coverage). These individual entities subscribing to a network engender small scale events. On the contrary, large scale events such as splits and merges are more frequent to occur in a dynamic environment. Due to mobility along with expansion, networks are likely to spatially overlap[1] and separate. Networking and especially routing require efficient and appropriate solutions. If we count upon heterogeneity, the dynamic nature of networks leverages particular difficult problems. We will detail a scheme based on AC principles that enables to merge wireless networks using distinct routing protocols to efficiently interoperate. We define it as an evidence that AC principles are pertinent in dynamic environments subject to merge phenomena. We give practical solutions to ensure our requirements are enforced.

The paper is organized as follows. Section 2 introduces a more general definition of merge and split related to the dynamics of communities. We formalize merges by defining their nature and implications. We then define merge and split as a general framework mainly related to interoperability issues. We give a set of requirements for protocols to be merge-compliant. Section 3 focuses on the implications of merging at the routing layer. We give a proof-of-concept of AC principles applied to the merge problem with two networks using distinct routing schemes, (i.e., AODV, DSR, and OLSR) and give a solution overview. We also tackle improvements to the proposed solution in order to scale with the increasing diversity of routing protocols. Finally, Section 4 discusses future research investigation and concludes this paper.

2 A General View of Merge

In the following, we mainly focus on merging of wireless networks and leave splitting for future investigation.

---

[1] Here we narrow merging to a physical overlapping but a more general definition is given in the following.

2.1   What Is a Community or Autonomic System

A general definition of a group is a number of network elements that share a common set of stable patterns of interactions. These interactions are essentially driven by social relations (e.g., meeting), involvement in a collaborative activity (e.g., P2P file sharing, work meeting, students' lecture, battling troops, emergency rescue teams) or with similar spatial patterns or simply geographic proximity (e.g., public transport users). These interactions – often correlated – can be represented by dynamic graphs of interactions in space and time; and mobility is only the visible part of these interactions. Note that this does not preclude a network element or AC system to belong to several groups.

2.2   Why Communities Merge and Split

Predicting patterns of interactions is a hard task given their dynamics. Some interactions are predictable while others are not. For example, social interactions are predictable such as regular meetings or workshops. Some mobility patterns are also predictable such as for users of public transports. The dynamics of a group can be classified in two types of events: small scale events such as node arrival, node departure, or node failure and large scale events such as splits and merges.

A general definition of merging is when two or more AC systems interact in order to collaborate whether spatially close or not given that a communication means is possible. The level of collaboration between these groups depends on their purposes' correlation. Two emergency teams following a similar goal (high correlation) will require to merge when meeting while two groups with different purposes, e.g., different WiFi operators spatially overlapping, might collaborate in order to interfere the least given the radio spectrum available or on the contrary offer roaming to their respective users. When merging the level of collaboration is reflected by the distinct groups' policy toward merging. The question is often to merge (high level of collaboration) or not to merge (low level of collaboration)? Depending on the negotiated level of collaboration, merge occurs at different layers (from the physical (PHY) to the application (APP) layer) and different time scales. For example, temporary splits may arise from a broken radio link due to the radio channel degradation or persist when mobility engenders sparse networks where network elements' radio coverage does not intersect.

2.3   Implications of Merging

Merges and splits depending on the level of collaboration is a source of conflict at all levels. When AC group merges, there are two great classes of conflicts:

- Heterogeneity of protocols from PHY layer to APP overlays,
- Resource driven conflicts, i.e., resource conflicts occurring at a given layer for homogeneous protocols. For example, the use of the same radio technology often leads to channel interferences. At the IP layer, merging requires to synchronize the addressing space of both networks in order to avoid conflicts.

2.4   Merging Requirements

Fundamental requirements of splits and merges are for AutoComm systems to keep what we call consistency. Consistency is the capacity for networks to keep their QoS and service level whatever small or large scale events occur at any time scale. Schemes dealing with merging and splitting must maintain or enhance the level of consistency by means of collaboration.

The second requirement is smoothness. Merges and splits are large scale events that can have a great impact on network performances. Smoothness is the capacity to cope with the smaller impact on the performances without disturbing the general QoS. In other words, we must design efficient, flexible solutions and what we can characterize as smooth split and merge solutions. Depending on the mobility pattern, merges and splits can be transient phenomena or on the contrary lead to stable situations. For example, current proposals tackling the problem of addressing in a merge environment assume an attraction/gravity mobility model where n wireless networks gather at a defined geographic location. This model leads to a permanent state where two networks spatially overlap. In this case, several schemes propose mechanisms to synchronize the address space in a coherent way so that no address conflict occurs. This can be done through flooding or other means. Nevertheless, with random mobility patterns merging leads to transient states where networks only cross by. In this case, re-addressing an entire network can be a sub-optimal solution.

The last requirement is efficiency. It requires from schemes to detect merge and split phenomena as quickly as possible and react appropriately.

With AC, going further than just detecting a phenomenon such as merging but by characterizing more deeply its nature (permanent, transient) will enable more scalable solutions to perform. These requirements require context-aware schemes.

2.5   Merge-Awareness

Merge-awareness is a kind of context-awareness or selfware. Context-awareness and what we define as merge-awareness enables to gather enough explicit information or if not available, to infer the underlying phenomenon occurring and take appropriate actions. For example, as explained before, mobility reflects one or several interactions a network element or AC is involved in. Characterizing mobility allows inferring the underlying interaction. In [11], the authors infer the will to merge of two wireless networks as shown in Fig. 1 by computing the relative velocity of both networks as a function of time. If this velocity is likely to converge toward zero, networks decide to merge since it reflects a tendency to effectively collaborate and leads to a stable state that will permit to optimize reconfiguration if needed. Other inference schemes carried out at a higher level study social interactions as an input [12], [13].

To enhance context-awareness, again if we borrow concepts from biological cells, AC systems require a memory similar to the immunological memory. The immune system and its memory allow efficient response to subsequent encounters with similar infections. The AC matching piece to infections is basically the environment context. Since situations (hence similar context) are likely to re-occur frequently given Zipf law, an environment context memory is required. As well, given the same reasons, merges and splits are to re-occur frequently. Keeping tracks of past merges and splits and recognizing an AC group that has partitioned in the past may enable an efficient re-merging thanks to a past shared context.

Solutions to cope with splits have been more tackled since splits are more predictable in their nature. Solutions are twofold, either they use a reactive approach i.e., detect the occurrence of splits and react appropriately by replicating what we call the patrimony in both splitting networks so that each networks is a duplicate (similar to bio-cell mitosis) or they use a proactive approach by periodically spatially replicating required patrimony in case of future split occurrences [14] [15]. The patrimony refers to the sum of all available services and information that are required by an AC group to still be autonomous in case of splits. The purpose of these schemes is to keep the consistency of both separating networks. There lies a trade-off between both approaches.

2.6   Interoperability

Merge is a very challenging issue in a dynamic environment. All layers are impacted by merging; from PHY to APP layers. Hence, merging has to cope efficiently with heterogeneity at all layers. As stated before, the AC paradigm will create autonomic systems using heterogeneous protocols. Similarly to biological systems that are defined by their fitness [5] as the ability to fit their local environment, routing must follow the same concepts. What we require from AC is to fit all situations. Routing must consider the group's purpose and the nature of the underlying phenomenon of the group's dynamic at small and large scale. This will require routing interoperability following the edicted requirements. We study such a case in the next section.

3   Proof-of-Concept

In this section, we give a proof-of-concept of AC principles applied to merging networks. We give an example that assists self-organization of network elements between AC groups using different routing protocols (routing protocol heterogeneity). We recall that in our vision, we consider AC as a means to federate existing solutions by enabling interoperability. This is a first step before fully compliant AC protocol design. This interoperability is subject to policy rules and requires specific function to sense the environment and bring context-awareness.

As stated before, merging requires being as smooth as possible. Since it is a large scale event, repercussion must be minimized for both merging AC groups.

Most ad hoc routing protocols are well suited for particular situations and hence are rigid. As said before, this evolution comes from the fact that routing protocols are a direct adaptation of wired routing protocols to the wireless world. Hence, the rigid nature of current ad hoc routing protocols leaves little space for adaptation. One solution we detail in the following is an AC daemon that enables a dynamic interoperability between rigid networks. Nevertheless, it is important to say that adaptive routing protocols such as ZRP are a first attempt to adapt the routing parameters to the sensed underlying environment.

3.1   Model and Hypothesis

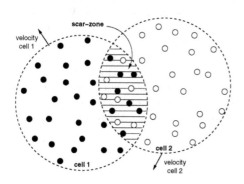

Fig. 1. Network cell model

In this paper, we define the concept of network cells. A cell, C, is the spatial region spanned by a set of nodes, N_C, willing to participate in some collaborate activity or having similar spatial behaviours. The evolution of a cell is progressive, i.e. new arriving nodes acquire the address allocation scheme and routing scheme from an existing cell member. The address assignment may be of any kind (conflict-free, conflict-detection, or best-effort). For example, nodes can acquire an address from their neighbours or randomly generate their addresses and verify their uniqueness using a duplicate address detection (DAD) [16] mechanism. For routing, we assume that due to the diversity of situations, nodes implement a set of existing protocols. However, since equipments have heterogeneous capabilities (i.e., processing, memory), they do not always implement the same set of routing protocols but it is likely that they will implement the most adopted ones. We believe that a restricted list of protocols will be supported by vendor equipments, some respecting standards and others proprietary implementations. When cells C_1 and C_2 are merging, the overlapping region $S_{C_1,C_2} = C_1 \cap C_2$ is called a scar-zone and is delimited by a scar-zone membrane (cf. Fig. 1). Nodes located in the scar-zone are called scar-zone nodes, S_C, while nodes outside the scar-zone, I_C, are named interior nodes. Depending on the respective mobility of the two cells, the scar-zone can evolve between a minimal overlapping where the two cells are interconnected via one radio link to a complete overlapping where the spatial extent of one cell is included in the other.

3.2   Routing Merge Requirements

We consider the case where cells C_1 and C_2 use heterogeneous routing protocols restricted to OLSR, AODV, and DSR. As described in section 2, we require merging to be as smooth as possible since it is a large scale event. The intuitive solution to the case we are dealing with would be to reconfigure entirely one of the two cells in order to have the same routing protocol. This requires all reconfigured nodes to implement the new routing protocol. Moreover, in a mobile

environment successive reconfiguration due to successive merging in a short lap
of time will lead to oscillations. Even if both cells use the same routing protocol (e.g., OLSR), synchronizing link-state (LS) tables might be sub-optimal if
merging is only transient. Besides, swapping to a different protocol will incur a
patrimony loss (on-going communications, routing states) and more importantly
break all Service Specific Routing (SSR) overlays relying on the previous routing
protocol. Recall that the choice of the physical routing protocol relies highly on
both the underlying environment and the requirements of the applications unless
both routing protocols can cohabit but this raises scalability issues. We believe
our scheme will benefit to situations of transient merging. These occur either
when networks just cross by (no stable overlapping, constant relative velocity
during merging) or during the transient phase occurring with the attraction mobility pattern (i.e., short period during which the relative velocity of merging
cells is non zero). We plead for a transient solution that will take effect until a permanent situation is detected and that will be able to choose the most
suited protocol given the context or find the appropriate parameters for adaptive
routing protocols.

3.3   Design

We briefly explain our approach here. The purpose is not to explain in detail the
mechanisms required for loop-free routing with heterogeneous ad hoc routing
protocols but to address the problems arising and give insights toward more
efficient and innovative solutions.

In order to achieve interoperability between merging cells, scar-zone nodes
must define their neighbourhood environment context i.e., the routing capabilities of their neighbours and the current protocol in use –the mother routing protocol. This is done with the Neighbourhood Routing Protocol Discovery
Protocol (NRPDP). For example, in Fig. 2 node X sends {*AODV*,DSR} to Y,
pointing out the routing protocols it supports and the current routing protocol
in use in its cell, indicated by *Protocol*. We supposed nodes or both cells to
have a common set of routing protocols but are not currently using the same one
when merge occurs. Depending on the neighbourhood context, specific interaction must be performed. Scar-zone nodes must either translate routing packets
or, if translation is not possible, nodes must execute appropriate neighbourhood
interaction. We define a new routing daemon, the Routing Translator Daemon
(RTD), that intercepts the I/O of routing control packets (requests, replies, and
updates) and given the context information provided by the NRPDP processes
these packets accordingly.

3.4   Application: AODV ↔ DSR

We give here an application of our framework for the interoperability case between AODV and DSR. Since AODV and DSR belong to the same family i.e.,
reactive routing protocols, a translation is only required. On the contrary, if we

required interoperability between a reactive protocol and a proactive protocol (e.g., OLSR) an other kind of interaction should have been used.

Consider the scenario shown in Fig. 2, where cell 1, C_1, runs AODV and cell 2, C_2, runs DSR. Consider also two nodes, A and B, with $A \in I_{C_1}$, $B \in I_{C_2}$ (i.e., $\{A, B\} \not\subset$ scar-zone). Here, we study how paths can be established in these cells in both ways, from A to B and vice-versa.

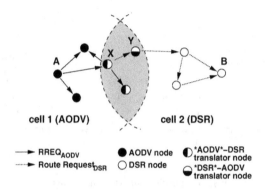

Fig. 2. Translation AODV $\leftrightarrow$ DSR

For the establishment of a path from A to B, $A{\rightarrow}B$, A floods a RREQ (Route Request). When a node in the scar-zone receives a request, here X, it translates the AODV RREQ into a DSR Route Request. This translation is associated with a new entry in a dedicated table maintained by the RTD. This entry will indicate that a translation will be required at the reception of the DSR Route Reply. When the DSR Route Reply is received, an AODV RREP is sent using the reverse route entry established by the initial RREQ. In order to avoid loops, the translation requires the use of the same request identifier, the same sequence numbers (d_{seq} and o_{seq}), and the precise translation of the number of hops. Since the sequence numbers of both protocols have different field sizes and to enable recursive translation, we use a hash function to associate AODV's RREQ sequence numbers with DSR's Route Request and add a new header dedicated to our RTD daemon. For the hop count, in the AODV header a hop count field represents the number of hops, while in DSR counting the number of concatenated IDs gives the hop number. Consulting the corresponding entry in the RTD table does this translation. As well, the correct association between routing control packets prevents recursive translations. For example, X receives an AODV RREQ from Y in reply to its original DSR Route Request sent to Y. By comparing the packet id and sequence numbers with the entry in its table, X detects this route request is generated in response to its original request.

Similarly, for the path $B{\rightarrow}A$, B floods a DSR Route Request that is translated into an AODV RREQ by scar-zone nodes. These scar-zone nodes will receive RREPs that will be translated back to DSR Route Replies.

3.5   Extensions to Our Model

Here, we relax some constraints on our hypothesis in order to obtain a more realistic model. In the previous model, we supposed all nodes to have similar capabilities at the routing layer referred as the routing capabilities or RC set. We supposed nodes to have a set of homogeneous routing protocols but now we consider that some nodes have smaller capabilities or different capabilities than others.

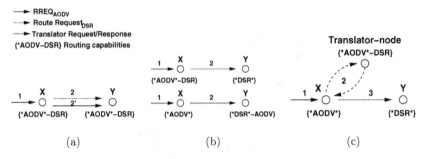

Fig. 3. Specific-context translations

Figure 3 lists all the possibilities for the previous case with a wider hypothesis on nodes' capabilities in terms of supported routing protocols. The different cases that occur will influence the reaction of our RTD given the neighbourhood information communicated by the NRPDP as the following:

- The first case, shown in Fig. 3(a) reflects the hypothesis of previous sub-sections where $RC_X \sim RC_Y$. Here, node X can either forward the AODV RREQ as it is or translate it into a DSR Route Request as long as the node that makes the translation updates its RTD table, i.e., $RC_X \sim RC_Y \rightarrow X_{translation} \lor Y_{translation}$.
- The second case, shown in Fig. 3(b) is when $RC_X \subset RC_Y \lor RC_Y \subset RC_X$. Here, the translation is done by the most capable node, i.e., $RC_Y \subset RC_X \rightarrow X_{translation}$ otherwise $Y_{translation}$.
- The last case, shown in Fig. 3(c) is the most complex to deal with. Here $RC_X \cap RC_Y = \{\}$. In this case, network entities require to find a node with more capabilities able to work out the translations. We call these nodes, translator nodes. We need these nodes to organize in an SSR overlay. Whenever a translation is required, one of these nodes works out the translation on behalf of the incapable node. How the overlay of most capable nodes is structured depends directly on the underlying routing protocol. In reactive protocols, nodes must use expanding ring search to find a translator node while in proactive approach where the topology is known, special entries can be added in the LS update packets or maintained by a new daemon dedicated to maintain the translator overlay. As shown in Fig.3(c), node X requests a translator-node which has a greater routing capability and thus

is able of performing the translation on X's behalf, i.e., $RC_X \cap RC_Y = \{\} \rightarrow Find_{translator}$ where $Find_{translator} \wedge$ *Reactive routing protocol* $\rightarrow$ *Expanding ring search* or $Find_{translator} \wedge$ *proactive routing protocol* $\rightarrow$ $Request_{SSR-translator-overlay}$.

Note that we have limited the cardinal of the capabilities set to 2, $|RC| = 2$. But, other more complex possibilities could enhance our approach. The last case requires extending our scheme with a SSR overlay dedicated to routing translation.

4   Conclusion

In this paper, we have addressed one of the future challenges networks will be faced to. We characterized splits and merges as large scale events that occur at different time scales with their causes and implications. We reviewed existing solutions and proposed yet simple but promising solutions for ad hoc routing interoperability. We have shown how AC can fully respond to challenges of merge and split in heterogeneous environment. We can draw several conclusions. First, if reactive/programmable approaches were used there would be no need of such a scheme. As we expressed before, reactive approaches will surely reappear for radio access technologies (PHY-MAC layer) with SDR [17] which enables a radio interface to be reconfigured by software. This will enable to swap from Bluetooth to IEEE 802.11 or ETSI HiperLAN, and to 3G and 4G. Nevertheless, we must take into account that reconfiguration can lead to sub-optimal solutions and initiate oscillations if splits and merges are to occur frequently. Second, our scheme will not scale with an increasing diversity of ad hoc routing protocols. With our designed scheme, $n(n-1)/2$ general translations rules are required and as much specific translation rules (cf. sub cases of extensions) with n as the cardinal of existing routing protocols. We believe our proposition will help in entering a new phase for routing. Our proposition has two opposite goals. On the one hand, respond to the urgent need of interoperability and in the other hand, our proposition is aimed at showing that interoperability is not always feasible and that innovations are still required in order to not reproduce errors of the past (e.g., ITU's Interworking units, OSI's internetworking (IDRP)). Since the tendency is toward protocol heterogeneity, we believe AC principles must enable protocol interoperability easily with more adaptable protocols. We are currently carrying out simulations using NS-2 in order to validate the suitability of our scheme under various mobility models.

References

1. Stoica, I., Adkins, D., Zhuang, S., Shenker, S., Surana, S.: Internet Indirection Infrastructure (i3). In: Proc. of the ACM SIGCOMM Conference (SIGCOMM'02), Pittsburg, PA, USA (2002) 73–88
2. Tschudin, C., Gold, R.: Network pointers. In: First Workshop on Hot Topics in Networks (HotNets-I), Princeton, NJ, USA (2002)

3. EURASIP: Special issue on UWB - State of the Art. In: Journal on Applied Signal Processing. EURASIP (2004)
4. Shafi, M., Gesbert, D., shan Shiu, S.D., Tranter, W.: Guest editorial: MIMO systems and applications. In: JSAC. Volume 21-3. IEEE (2003)
5. MacKay, D.: Information Theory, Inference, and Learning Algorithms. Cambridge University Press (2003)
6. Clausen, T., Jacquet, P.: Optimized Link State Routing Protocol OLSR. RFC (2003) IETF RFC-3626.
7. Johnson, D., Maltz, D.: Dynamic Source Routing in Ad Hoc Wireless Networks. In: Mobile Computing. Volume 353. Kluwer Academic Publishers (1996)
8. Perkins, C., Royer, E.: Ad-hoc On-Demand Distance Vector Routing. In: Proc. of MILCOM (MILCOM'97). (1997)
9. Haas, Z.: A New Routing Protocol for the Reconfigurable Wireless Networks. In: Proc. of the IEEE International Conference on Universal Personal Communications, San Diego, CA, USA (1997)
10. Ramasubramanian, V., Haas, Z., Sirer, E.: SHARP: A Hybrid Adaptive Routing Protocol in Mobile Ad Hoc Networks. In: Proc. of the ACM International Symposium on Mobile Ad Hoc Networking and Computing (MobiHoc'03), Annapolis, MD, USA (2003)
11. Legendre, F., Amorim, M., Fdida, S.: Implicit Merging of Overlapping Spontaneous Networks. In: Proc. of the IEEE Vehicular Technology Conference 2004-Fall (VTC'04), Los Angeles, CA, USA (2004)
12. Wang, B., Bodily, J., Gupta, S.: Supporting Persistent Social Groups in Ubiquitous Computing Environments Using Context-Aware Ephemeral Group Service. In: Proc. of 2nd IEEE International Conference on Pervasive Computing and Communications (PerCom'04), Orlando, FL, USA (2004)
13. Kurvinen, E., Oulasvirta, A.: Towards Socially Aware Pervasive Computing: A Turntaking Approach. In: Proc. of 2nd IEEE International Conference on Pervasive Computing and Communications (PerCom'04), Orlando, FL, USA (2004)
14. Wang, K., Li, B.: Efficient and Guaranteed Service Coverage in Partitionable Mobile Ad Hoc Networks. In: Proc. of the 21st IEEE Infocom (Infocom'02), New-York, USA (2002)
15. Hanna, K., Levine, B., Manmatha, R.: Mobile Distributed Information Retrieval for Highly-Partitioned Networks. In: Proc. of the 11th IEEE International Conference on Network Protocols (ICNP'03), Atlanta, Georgia, USA (2003)
16. Nesargi, S., Prakash, R.: MANETconf: Configuration of Hosts in a Mobile Ad Hoc Network. In: Proc. of the 21st IEEE Infocom (Infocom'02), New York, USA (2002)
17. JSAC-IEEE: Special issue on Software Radios. In: Mobile Computing. Volume 17-4. IEEE (1999)

Policy Interoperability and Network Autonomics

Shane Magrath[1], Robin Braun[1], and Fernando Cuervo[2]

[1] University of Technology, Sydney,
Broadway, Australia
shane.magrath@uts.edu.au
[2] Research and Innovation, Alcatel Canada
600 March Rd, Ontario Canada K2K 2E6,
fernando.cuervo@alcatel.com

Abstract. Autonomic behaviours in network operations will alleviate much of the labour intensive and error prone interventions of today's complex networks. The Service Provider must be able to manage the infrastructure and services at an abstract level, focusing on what the desired behaviour should be rather than how it might be specifically achieved. Policy-Based Network Management (PBNM) appears as one of the leading mechanisms to describe desired behaviours and abstract the programmability of an autonomic network infrastructure to the Service Provider. For massive-scale and complex networks, the current understanding of the Higher Level to Lower Level (HL→LL) refinement process commonly used in PBNM today is not completely effective. One problem encountered is the need to provide a bind mechanism between Higher Level and Lower Level policy specifications such that cross-layer policy requests in the policy continuum can be made by lower policy layers in a dynamic policy refinement cycle (LL→HL→LL). In this paper, we illustrate the problem with a policy-based simple admission control (SAC) application. We then show that policy specifications with a join operator ($\bowtie$) simplify the SAC specification. We also investigate the performance considerations of this enhancement in Internet size applications. Our future goal is to provide a policy inference engine that can support complex specifications appropriate for PBNM systems that support autonomic behaviours in large networks, made of Network elements with realistic memory and processing constraints.

1   Introduction

There can be no question that infrastructure and services are harder to manage now than they were perhaps five years ago. In that time, technology has improved, customer expectations have risen, services have become more complex, the weight of legacy infrastructure heavier and the collision between traditional Telecommunications applications (*voice*) and Enterprise technology (TCP/IP) have threatened the sustainability of conventional market models for Carriers and Service Providers.

Telecommunication systems management is *complex*. Conventional management architectures and standards have proven inadequate when faced with new

M. Smirnov (Ed.): WAC 2004, LNCS 3457, pp. 25–43, 2005.

sets of complicating requirements - pervasive operational security, differentiated service level agreements and a plethora of Next Generation Services. One interesting reason is that these legacy management architectures maintain the semantic locus in the *individual* elements that comprise the domain of managed objects. Data is collected from *each* device, and stored centrally. Information about service status is attempted to be re-constructed from a collating of the properties of the *individuals* from the network. This is analogous to considering a person's health as a function of "simply" probing the state of each of the body's cells and collating the findings - an ultimately ineffective approach since it can not describe the state of the higher physiological functions that *emerege* when aggregates of cells interwork to give rise to a new system functions.

In this context, policy-based management architectures have been considered as viable and necessary part of these new management frameworks ([1, 2, 3, 4]). Service Providers need to be freed to manage their systems at a higher level of abstraction than the mere technology configuration. They need to consider and specify the *business* requirements of the *applications* that comprise the *services* being operated. With a proper understanding of the *roles* that comprise the service operations, the Service Provider can formulate HL policies appropriate for a PBNM system that link *business* requirements with *technology* configuration - a level of interoperability that has previously not been achieved.

However, the HL→LL refinement process is not always sufficient for the needs of complex networks and services. Typically, the managed objects within the domain will encounter situations not covered by their present configuration. These devices need a means to determine appropriate behaviour when faced by these conditions. By referring the request to a policy server that is authoritative and capable of interpreting the request against the HL policies, an appropriate LL policy can be identified and deployed to the device. This LL→HL→LL cycle can be problematic. It implies that the policy server is able to bi-directionally refine HL and LL policies in real-time. This further implies a mechanism whereby the policy server can provide a binding between the LL and HL policies. Unfortunately, most of the current policy specification approaches and languages do not innately have this ability.

In this paper, we consider the value of a *join* operator in a policy specification. The join operator allows a linkage to be achieved between HL and LL policy information. This can greatly simplify policy specification for the LL→HL→LL requirement. Furthermore, the join operation is required to perform at the level appropriate for these real-time systems where servicing massive-scale applications involve transaction rates that are measured in thousands of events per second. For autonomic system architectures involving a centralised PBNM system these issues need to be considered.

Our paper proceeds as follows. Section Two considers the current state of the research in this area. Section Three presents a scenario involving a simple access control application (SAC) and examines the issues presented by this application. Section Four presents a contribution towards alleviating the problem identified by

SAC and a consideration of the issues raised by the proposed solution. We finish with a summary of our findings and an outline of future work to be undertaken.

2   Previous Work

We consider the literature from the perspectives of:

- policy semantics, specifications and languages
- policy refinement

2.1   Policy Semantics

A bit of space will be taken to outline the semantics of policy seen in the literature since a characterisation of these systems is needed here.

Within the literature, the term "policy" generally means an *administrative rule* - that is, a *declarative* statement of requirement. More specifically though, the semantics of policy vary slightly across the literature. The IETF [5] defines policy as follows:

Policy

"Policy" can be defined from two perspectives:

- A definite goal, course or method of action to guide and determine present and future decisions. "Policies" are implemented or executed within a particular context (such as policies defined within a business unit).

- Policies as a set of rules to administer, manage, and control access to network resources [RFC3060].

Note that these two views are not contradictory since individual rules may be defined in support of business goals.

As can be seen from the IETF's definition, policy involves notions of "context" and abstraction. More helpfully, Verma [6, 7, 8, 9] has drawn the distinction between high-level policies and low-level policies. High-level policies are used to express "business-level" rules. Low-level policies are used to express "technology-level" rules.

Again, the IETF [5] also develops the notion that policies have varying levels of abstraction:

Policy Abstraction

Policy can be represented at different levels, ranging from business goals to device-specific configuration parameters. Translation between different levels of "abstraction" may require information other than policy, such as network and host parameter configuration and capabilities. Various documents and implementations may specify explicit levels of abstraction. However, these do not necessarily correspond to distinct processing entities or the complete set of levels in all environments. (See also "configuration" and "policy translation".)

Abstraction is an important concern in autonomic systems engineering because it allows us to describe and be concerned only for the important functions of the required abstracted autonomic behaviour.

The most significant body of research contributing to PBNM has been undertaken at Imperial College under Morris Sloman. A summative work in [1] describes their definition of policy to include "*types*". They classify policy as either being typed as:

- Authorization - related to the permissions that a "*domain subject*" can perform,
- Delegation - related to the ability of a domain subject to delegate its privileges,
- Obligation - related to the actions a domain subject must perform in response to conditions and events,
- Refrain - related to the actions a domain subject must refrain from performing on "*targets*" in the domain.
- Composite - a grouping of the more basic types described above for administrative reasons.

Closely associated with policy is the concept of "*roles*". Roles are generally used as a container for policies. That is, a role can contain a collection of policies. Moreover, roles express the rights, duties and obligations of a position or function. The role concepts have been developed by Sloman et al in [1, 10, 2].

Sloman et. al. also add to the concepts of policy and PBNM by incorporating the concept of "*Domains*" into the semantics ([11, 12, 13]). Domains are a collection of managed objects that are under one administrative control and are related by "subject". That is, those objects that are part of the same policy application space may be collected together in a domain. Policies may operate on the entire set of domain objects or a sub-set defined by some selection criteria. For example, a Service Provider may place all DIFFSERV edge routers in the same domain. A policy may then be authored such that the *scope* of the policy is limited to a sub-domain of those routers - for example, all DIFFSERV edge routers located in Victoria (a subset of Australian routers) should authenticate Operations Support staff who wish to log on to the router via the Victorian RADIUS server.

A common thread to policy definition is the importance placed on the concept of "*events*". Events are a signal from the management environment that a possible state-change to the Domain has occurred. Events are used to trigger policy evaluation ([14]). In contrast to most other approaches, the IETF COPS formulation has no precise operationally explicit syntax for event management. However, there is an implicit concept of events in the QoS policy applications where packet arrival events are used to trigger the appropriate evaluation and marking of the packets as specified by the QoS PIB.

Several contributions have been made to the specification of policy *languages*. Two significant contributions are:

Ponder   The formal specification from Imperial College ([15, 14]). Ponder is a declarative, object-oriented language that supports events, constraints, roles, templates and other useful language features.

$\mathcal{PDL}$ Policy Description Language ($\mathcal{PDL}$) is from Bell Labs ([1, 16, 17]). $\mathcal{PDL}$ is a declarative event-condition-action language originally developed for specifying network management policies.

Ponder. A whimsical and hopefully self-explanatory illustration is provided:

```
inst oblig /Policies/HomeLandSecurityPolicies {
    on        Event(TerroristAction, Hostage) ;
    subject   /Government/MI5 ;
    target    t = /Agents/Agent007 ;
    do        t.CaptureTerrorists(TerroristAction)->
              t.RescueTheGirl(TerroristAction, Hostage) ->
              t.SaveTheWorld(TerroristAction) ;
    when t.isNotInBed() ;
}
```

$\mathcal{PDL}$. Policy Definition Language ($\mathcal{PDL}$) ([16, 17]), like Ponder, is an event-condition-action (ECA) declarative language though it does not have all the features that Ponder has. $\mathcal{PDL}$ is reviewed by Sloman in [1]. $\mathcal{PDL}$ was originally developed for the specification of network management policies.

$\mathcal{PDL}$ policies consist of two types of expressions:

- policy *rule* propositions of the form :
  `<event> causes <action> if <condition>`
- policy *defined event* propositions of the form :
  `<event> triggers pde(`$M_1 = T_1, ..., M_k = T_k$`)`
  `if <condition>`

The policy rule is the conventional ECA specification of a rule. The policy defined event read "if the *event* occurs and the *condition* is satisfied then the policy defined *event* is triggered". $\mathcal{PDL}$ supports a basic event calculus for causal specification purposes. $\mathcal{PDL}$ does not support the notion of "roles", nor does it have a concept of "domains". These are serious weaknesses to have in a *generalised* Pbnm system.

Despite the simplicity of the language, $\mathcal{PDL}$ has shown itself capable in managing a range of network management tasks involving telecommunications switch products ([1]).

2.2   Policy Refinement and Inter-operability

Policy refinement is concerned with the process of mapping a set of HL policies to a set of LL policies. Bandara ([14]) considers refinement as having three requirements: Correctness, Consistency and Minimality. Verma identifies the correctness requirements for successful refinement by describing the process as a

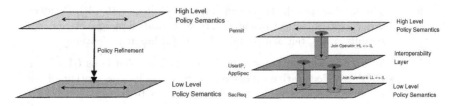

Fig. 1. (a) Policy Refinement (b) Policy Interoperability with joins via Intermediate Layer

consideration of translation, bounds, relation, consistency, dominance and feasibility checking ([6]). Kanada considers refinement more critically in [18,19] in terms of the significant problems of optimally refining HL policies to LL policies involving policy division and fusion.

In contrast to refinement, we introduce *inter-operability* (see Figure 1). Refinement is concerned with the one directional mapping: HL→LL. Inter-operability is the bi-directional mapping: HL ↔ LL. The refinement process occurs *before* policies are operationally deployed. Inter-operability occurs *as part of* the operational deployment. The function of inter-operability mapping is to allow LL policies at run-time to dynamically refer to their HL parents as the need arises. We will see that interoperability requires the presence of an interoperability layer (IL) that mediates between HL and LL representations.

In the most general case where there exists a progression of abstraction layers (the "policy continuum" [20]), the HL ↔ LL mediated by an IL remains useful as a fundamental pattern. By cascading the inter-operability model, a more general abstracted policy continuum can be achieved - "one man's HL is another man's LL" so to speak. It is quite common for autonomic systems to contain quite a sizable stack of abstraction layers to provide the final functional service, so the ability for inter-operability to be cascaded in order to maintain the interchange between layers is reassuring. In contrast, policy refinement presents some problems to autonomics because each downward refinement from one HL to the next LL further distances the final operations from the true functional intent of the uppermost policy management layer. Without the ability for the LL to interact with the upper layers, policy refinement only provides a partial mechanism for autonomic operations.

3   Theory

We begin our consideration of the policy refinement and inter-operability problem with a problem scenario. We imagine a *large* network (carrier-scale) of users who enjoy the benefits of *individually* tailored service levels differentiating the quality of service their *applications* receive (see Figure 2). Moreover, the network provides a rich set of features such as broadband mobility, ubiquitous service access, and continuous context sensitive display to multi-modal terminals.

In this scenario, it is the *access* network that needs to perform the necessary access controls and QoS management to enforce the service level requirements.

Moreover, in this environment it does not make sense necessarily to *pre-provision* the edge devices with complete and specific LL policy configuration: there are a lot of users and they are mobile. Pre-provisioning policy to the edge consumes resources, with consequences if poorly deployed:

- large LL policy tables occupy more memory and therefore slow search times;
- mobility means that not all the required LL policies are in the tables. Moreover, there may exist LL policies in the tables that won't be used in practice;
- the policy server is committed to maintaining the state of deployed LL policies with no benefit if the policies are inappropriately deployed.

By deferring the deployment of LL policies until it becomes clear of the location of the user and their *contextual* requirements, a better match between policies usefully deployed and the resources consumed is made. This late binding of policies can benefit the system by mediating the effects of poor policy deployment, as argued in [21].

If we restrict our consideration to *simple access control* with deferred policy provisioning, then we require each edge device to ask the policy server what to do when it detects a new session flow (a "context request") that involves a previously unknown user, or a previously unencountered combination of source/destination and application specifications for the new session.

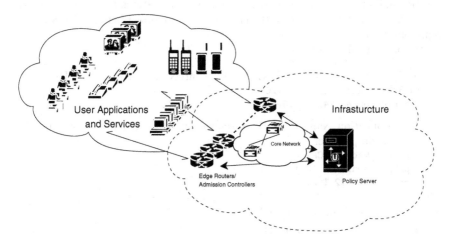

Fig. 2. Simple Access Control in a Network

In this case, the edge device informs the policy server by sending an *event* containing the tuple:

```
{Sac_Request SrcAddr SrcPort DestAddr DestPort Protocol SessionID}
```

This tuple contains the minimum information specifying a new session flow in the network. It is the LL specification of information appropriate for the technology plane of systems management.

However, the Administrator is happily oblivious to these specifics. He relates to the PBNM system through the specification of HL policies appropriate to the requirements of the business. In the SAC case, the Sys Admin has previously made known to the system the following tuples of information:

```
{Permit Shane Web}
{Permit Shane email}
{Permit Shane SSH}
{Permit Leanne Web}
{Permit Leanne email}
```

where each tuple is of the format

```
{Permit UserName ApplicationName}
```

Here, the user SHANE is permitted to use web, email and SSH applications. Implicitly, any application not in the list is denied access to the network. Similarly, user LEANNE is permitted to use only web and email. Here the requirements are analogous to a firewall. However, *every* edge device in the domain performs the firewall enforcement function.

If this were the only system information available, there is no possibility for the PBNM system to effect the meaning of the HL policies at the LL operational level. Additional information, from ad hoc sources, that serves to bind the HL and LL representations is required. These relations form an *Interoperability Layer* (IL) and facilitate the cross-domain mapping between LL and HL representations.

We need two IL relations:

- A User Name $\Longleftrightarrow$ IP Address binding, sourced from RADIUS, or DHCP :
  ```
  {UserIP Shane 138.25.41.126}
  {UserIP Leanne 142.53.16.7}
  ```

- An Application Name $\Longleftrightarrow$ IP Specification binding, made known by the application provider :
  ```
  {AppSpec Web 80 TCP}
  {AppSpec email 25 TCP}
  {AppSpec SSH 22 TCP}
  ```

where each tuple is of the format[3]

```
{AppSpec Name Port Protocol}
```

It is now conceivable that the policy server can determine what to do in response to SAC_REQUEST events. For instance, if the server were to receive:

---

[3] In reality, the specification of applications in terms of protocol and port numbers are more complicated than this since more than one tuple may be required and the tuples may be dynamic. However, this is simple access control after all.

Algorithm 1. Policy Rule for SAC (expressed in the Jive! language)

```
policy SAC_Policy {
    condition {
        event Sac_Request : (srcAddr * * destPort protocol SessionID}
              UserIP      : (userName srcAddr}
              AppSpec     : (appName destPort protocol}
              Permit      : (userName appName}
    }
    action {
      main {
        Send(Sac_Response, SrcRouter, SessionID, Permit);
      }
      default{
        Send(Sac_Response, SrcRouter, SessionID, Deny);
      }
    }
}
```

{Sac_Request 138.25.41.126 1078 204.32.45.61 80 TCP}

it should reply with a POSITIVE authorisation.

However, to do so involves the server in several join operations:

− It has to perform a LL→HL resolution of the source address (138.25.41.126 →SHANE);
− It has to perform a LL→HL resolution of the application (80/TCP →WEB)
− It needs to determine if the request is permitted (SHANE + WEB →PERMIT)
− It finally needs to resolve the HL policy requirement to a LL deployable specification:

{Sac_Response SrcRouter SessionID Permit}

This simple application can be specified with the policy rule in Algorithm 1., using the Jive! language we have developed for experimenting with these systems.

The *condition* clause of this policy involves a syntax similar to conventional rule-based production systems. Join operations are identified by a "¡join_name¿" syntax on the right hand side of the ":" operator. An isolated "*" signifies a "don't care" conditional match for that particular field/attribute. The policy also includes a *default* action clause that allows for the efficient handling of requests that *fail* to cause the condition clause to evaluate to TRUE.

The most interesting feature of the policy rule is the *condition* clause. The condition clause defines the pattern matching relationship between the HL and LL data as well as the more general constraints of that relationship. By taking the set of set of data and looking for *all* combinations that can satisfy the condition clause pattern constraints, a set of activations that can be executed is achieved. Moreover, in this example, the use of the join operations fulfill the needs of the interoperability requirements.

This simple example provides support that the join operator is an elegant and effective means for dynamic policy refinement and interoperability *between* HL business rules and LL technology configuration. This policy has achieved the valuable goal of enabling the system to inter-operate between HL and LL levels of abstraction and simultaneously maintain the very nice separation of concerns relating to the specification of HL and LL information.

However, most of the current examples of policy languages and specifications do not support this operator. This is not necessarily an oversight. PBNM specifications originally developed to service a specific need: the management of QoS services, and DIFFSERV in particular. The requirements here were to provide a PBNM framework that can operate in real-time by providing LL policy configuration in a form appropriate for the domain of managed devices, namely *tables* of a *limited* range of *typed* information that devices such as routers and switches could interpret efficiently. In this context, keeping policy free of join semantics is conducive to the QoS management problem. It does however, make things difficult for dynamic HL↔LL interoperability.

To be considered is the question of *efficiently* evaluating the join in real-time. Despite the apparent simplicity of the SAC application, as the *number* of tuples grow, the number of candidate matches that need to be considered by the rule also grows exponentially. This presents a tension between requiring the ability to support HL policy specification that is fully interoperable with the LL specifications, and the certain need for maintaining system throughput performance at very high transaction rates and low round-trip latency times.

We proceed to consider these issues.

4   Considerations

4.1   Action Clause

One observation that can be made is that a similar effect of the join can be procedurally achieved as part of the *action* clause of the policy. That is, if we restrict the condition clause to just the SAC_REQUEST event specification, then the remaining information can be determined as a series of functional lookups to a directory, or a location service, etc as part of the action clause, an approach sometimes seen.

There are a few issues with this:

1. Functional lookups require the existence of the functions to perform them. These functions are either to be made available as libraries as part of the policy language, or the Sys Admin would need to develop them.
2. The question exists whether the sort/search/select operations required as part of the lookup leads to best performance of the policy server.
3. If every interesting event that is raised triggers a positive evaluation in the condition clause, only to be later discarded by further constraints in the action clause, where was the benefit? Moreover, by raising an abortive activation, other activations in the set that contend for service risk delayed resourcing.

4. Good design would suggest that the requirements are best served by maximising the necessary constraints on the condition clause leaving the action clause to be as specific and productive as possible. This approach is consistent with the very event-condition-action character of the system.

4.2   A More Formal Consideration

We will make some observations about the policy computational complexity by formalising the description of SAC.

We first define the set AS as the *activation set*. It is the set of all instances of *current* rule activations that have yet to be serviced.

We define the *system state* as a series of *relations* on the data known to the system. For example, in the SAC example, data about users is made known through the relation schema:

$$UserIP = \{UserName, IPAddr\} \tag{1}$$

Similarly,

$$
\begin{aligned}
Apps &= \{AppName, Port, Protocol\} \\
Permits &= \{UserName, AppName\} \\
SacReq &= \{SrcAddr, SrcPort, DestAddr, DestPort, Protocol\} \quad (2)
\end{aligned}
$$

We are particularly interested in the effects of the *SacReq* event on the activation set:

$$AS = AS \cup Rules(SacReq) \tag{3}$$

where

$$
\begin{aligned}
Rules(SacReq) &= R_1(SacReq) \cup R_2(SacReq) \\
&\cup ... \cup R_n(SacReq)
\end{aligned} \tag{4}
$$

and R_i is rule i in the system.

This expresses the idea that a *single* event may be responsible for *multiple* activations as more than a single rule may be satisfied by the event. This representation is particularly relevant when the underlying inferencing engine does not support "default action" semantics. See Appendix One for a short discussion.

For the next section, it helps to know that the join operator can be defined as:

$$A_a \bowtie_b B = \sigma_{a=b}(A \times B) \tag{5}$$

As an example:

if

$$A = \{a, b\}$$
$$B = \{1, 2\} \tag{6}$$

then

$$A \times B = \{(a, 1), (a, 2), (b, 1), (b, 2)\}$$
$$\therefore \sigma_{a=b}(A \times B) = \emptyset$$
$$\therefore A_a \bowtie_b B = \emptyset \tag{7}$$

We now consider the rule from Algorithm 1.:

$$AS = AS \cup SAC\_Policy(SacReq) \tag{8}$$

Procedurally, we express the rule predicate as a process of relational refinements using relational algebra:

$$J1 = UserIP_{IpAddr} \bowtie_{SrcAddr} SacReq$$
$$S1 = \sigma_{SacReq.SrcAddr}(J1)$$
$$J2 = Apps \underset{\substack{port \\ protocol}}{\bowtie} \underset{\substack{port \\ protocol}}{} SacReq \tag{9}$$
$$S2 = \sigma_{\substack{SacReq.Port\ AND \\ SacReq.Protocol}}(J2)$$
$$J3 = Permits_{UserName} \bowtie_{UserName} S1$$
$$J4 = J3_{AppName} \bowtie_{AppName} S2$$
$$AS = AS \cup J4$$

Now, if $|J4| = 0$ then the condition clause is *not* satisfied and the *default* action is added to the activation set. Otherwise, each tuple within $J4$ is added to the activation set for *executor* scheduling. When will $|J4| = 0$? When the SacReq event presents a context (that is, set of field values) that:

1. can *not* be mapped into the Intermediate Layer (IL) - (the "unknown application" or "unknown user" case)
2. *can* be mapped to the IL *but* can *not* be mapped from the IL to the HL - (the "no permission" case) (see figure 1).

If either of these two conditions hold, then $|J4| = 0$.

In terms of computational complexity, since the join operator can be defined as:

$$A_a \bowtie_b B = \sigma_{a=b}(A \times B) \tag{10}$$

So, if $|A| = N$ and $|B| = M$ then

$$|A_a \bowtie_b B| = |\sigma_{a=b}(A \times B)| \tag{11}$$
$$\leq M \times N$$

That is to say, the join operator is performed by :

1. taking the Cartesian product of the two relations (forming a new relation whose cardinality is $M \times N$),
2. selecting tuples from this intermediate relation that satisfies the join condition.

In this way, we can see that joining two *large* relations, there can be a substantial computational cost proportional to the product of their cardinalities. Within the SAC application, we should reasonably expect:

$$| \; Permits \; | \gg | \; UserIP \; | \gg | \; Apps \; | \gg 1$$

Given this, one might expect $J3$ to be the most expensive operation in the procedure.

A few observations can be made:

- the algebraic procedure above for the SAC_Policy is *not* unique. That is, there are other equivalent derivations that are mathematically identical in the final result. However, they are *not* cost identical. This means in practice we would wish to *optimise* the method for computationally determining the SAC_Policy result. The basis for this optimisation follows from the following observation.
- there is much to be gained to being as *selective* as possible *early* in the pattern matching process. By reducing the cardinalities of the intermediate relations as early as possible, the join operations become more efficient in both memory and time requirements;

4.3   Experiment

Design. We built a prototype PBNM system ("Step") that supports Join semantics for the purpose of testing, amongst other things, the performance qualities of the system against increasing domain size. As stated earlier, it is a requirement of all Service Provider PBNM systems to adequately perform for domain sizes consistent with those found in massive-scale Service Provider networks.

With reference to Figure 3, the experimental system consisted of three main components:

- GENEVA: a configurable Elvin Event Generator for producing the SAC_REQUEST events (developed by us),
- Elvin Server: the Elvin content-based routing server ([22] [4]),
- STEP: The policy enforcement point that we developed.

The system was established on the university's research computing cluster (Orion). Each of the machines in the system is composed of a 3.0GHz Pentium 4 with 800MHz FSB and 2GB 400MHz DDR-RAM and runs Red Hat Linux 8.0.

---

[4] Refer HTTP://elvin.dstc.com/

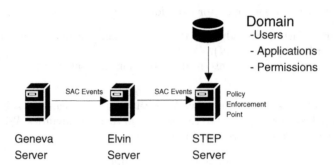

Fig. 3. Experiment Configuration

For each of the runs, the GENEVA application was configured to produced 20,000 SAC events at the rate of 40 events per second distributed negative exponentially to emulate the stochastics of large group behaviour. The field contents of each SAC event is randomised in such a way to range over the entire domain dimensionality. The STEP sub-system receives and enqueues the SAC_REQUEST events from the Elvin server. At the heart of STEP is an externally sourced rule inferencing engine (Jess: Java Expert System Shell[5]) that implements the Rete algorithm [23]. This algorithm performs the computation of Join operations and is the typical algorithm found in most commercial and academic inferencing engines. We encoded the SAC application into STEP so that it would dequeue the SAC events and take appropriate action (permit or deny) according to the policy. We instrumented STEP in order to measure the performance of the inferencing sub-system under increasing domain sizes. This data was captured for each experimental run that we performed.

The reported results are for a series of runs consisting of the number of known applications held constant at 10, and a fixed Permit cardinality of 40% of UserIP x Apps. The free variable is the number of Users the system knows about. The runs consists of User populations of 10, 100, 1K, 10K, 100K, 1M users.

Results and Discussion. Figure 4 reports the throughput characteristics of the system that was determined from the experimental runs described above. The results are not encouraging for the massive-scale applications envisaged by a Service Provider. For even moderately sized domains, the performance of the system deteriorates significantly reaching a minimum of around 20 events per second throughput. Service Providers need to maintain system throughput of the order of thousands of events per second for massive domains. We are several orders of magnitude below the requirement.

As indicated in Section 4.2, this result is not surprising because of the multiplicative effects observed in the Cartesian products comprising J4 (via J3 and J1). The

---

[5] Refer to HTTP://herzberg.ca.sandia.gov/jess/

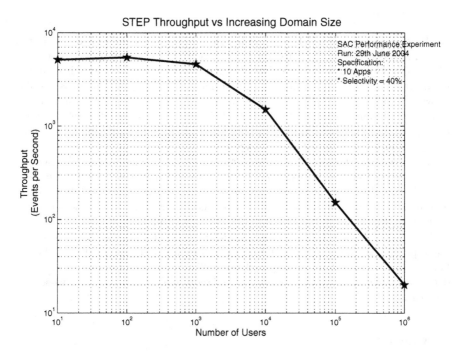

Fig. 4. Performance Results for STEP

Rete algorithm, as a finite-differencing algorithm, maintains increasingly large data structures commensurate with increases in domain size. Whilst it is not surprising that performance diminishes with domain size, the question then becomes how do we maintain performance *and* the use of the Join semantics which is so useful for autonomic and policy-based network management. This is our future work. The logical next step is to try other algorithms besides Rete, such as TREAT and Matchbox ([24, 25]). This is useful, necessary work however any purely centralised architecture will ultimately be defeated by sheer size. Ultimate improvements will largely be made through increased processing speed (Moore's Law) and improvements in compiler optimisation techniques.

An obvious alternative is a more distributed approach such as may be achieved by a multi-agent system, however our feeling is that the impact of inter-agent communications may defeat the advantages of the distribution. An interesting alternative, and certainly more consistent with physiological and biological autonomic systems, are the *swarm* algorithms for performing task allocation and resource distribution ([26, 27]). Their main advantage is their *lack* of inter-agent communication, and robust ability to *adapt* to changing environments. However, the engineering of such systems is still far from mature so this forms another line of development in our research.

5   Conclusion

We have established that LL→HL→LL policy interoperability provides significant Administration benefits to complex network management. The concept of
an interoperability layer that mediates LL and HL layers of abstraction is seen to
be an important component for the autonomic management of systems since it
allows the bidirectional policy interaction between the layers during system run
time without Administrator intervention. This is in contrast to policy refinement
approaches that seek to "compile" policies from a HL representation into a LL
specification prior to operational deployment.

We have also established the need for a high performance policy inferencing
engine that can service the needs of massive-scale real-time applications found in
large Service Provider networks. The use of standard algorithms for inferencing
may not be the best choice for the specific needs of Service Providers and the
type of real-time policy applications they may wish to run.

Our future work consists of developing a set of benchmark real-time policy applications that are relevant to Service Providers in general. Using these
benchmarks we expect to develop and test different inferencing algorithms and
determine which may best fit the operational requirements of these massive-scale
applications.

Acknowledgment

The authors also wish to thank Alcatel for their support and interest in this
research.

References

1. Sloman, M., Lupu, E.: Security and Management Policy Specification. IEEE
   Network **2** (2002) 10–19
2. Lupu, E., Milosevic, Z., Sloman, M.: Use of Roles and Policies for Specifying
   and Managing a Virtual Enterprise. In: Research Issues on Data Engineering:
   Information Technology for Virtual Enterprises, 1999. RIDE-VE '99. Proceedings.,
   Ninth International Workshop on. (1999) 72–79
3. Strassner, J.: Policy-Based Network Management - Solutions for the Next Generation. Morgan-Kaufmann (2003)
4. TMF: The NGOSS Technology Neutral Architecture Specification V3.0. Technical
   Report TMF053, TMF (2003)
5. Westerinen, A., Schnizlein, J., Strassner, J., Scherling, M., Quinn, B., Herzog, S.,
   Huynh, A., Carlson, M., Perry, J., Waldbusser, S.: Terminology for Policy-Based
   Management. Informational RFC 3198, IETF, Network Working Group (2001)
6. Verma, D.: Simplifying Network Administration Using Policy-Based Management.
   IEEE Network **16** (2002) 20–26
7. Rajan, R., Verma, D., Kamat, S., Felstaine, E., Herzog, S.: A Policy Framework
   for Integrated and Differentiated Services in the Internet. IEEE Network **13** (1999)
   36–41

8. Verma, D.C.: Policy-Based Networking: Architecture and Algorithms. New Riders (2001)

9. Verma, D., Beigi, M., Jennings, R.: Policy Based SLA Management in Enterprise Networks. Lecture Notes in Computer Science **1995** (2001) 137–??

10. Lupu, E., Sloman, M.: A Policy Based Role Object Model. In: Enterprise Distributed Object Computing Workshop [1997]. EDOC '97. Proceedings. First International. (1997) 36–47

11. Damianou, N., Dulay, N., Lupu, E., Sloman, M., Tonouchi, T.: Tools for Domain-Based Policy Management of Distributed Systems. In: Network Operations and Management Symposium, 2002. NOMS 2002. 2002 IEEE/IFIP. (2002) 203–217

12. Robinson, D., Sloman, M.: Domains: A New Approach to Distributed System Management. In: Distributed Computing Systems in the 1990s, 1988. Proceedings., Workshop on the Future Trends of. (1988) 154–163

13. Sloman, M., Magee, J., Twidle, K., Kramer, J.: An Architecture for Managing Distributed Systems. In: Distributed Computing Systems, 1993., Proceedings of the Fourth Workshop on Future Trends of. (1993) 40–46

14. Bandara, A.K., Lupu, E., Russo, A.: Using Event Calculus to Formalise Policy Specification and Analysis. In: IEEE Workshop on Policies for Distributed Systems and Networks (Policy 2003). (2003)

15. Damianou, N.: A Policy Framework for Management of Distributed Systems. PhD thesis, Department of Computing, Imperial College, London (2002)

16. Bhatia, R., Lobo, J., Kohli, M.: Policy Evaluation for Network Management. In: INFOCOM 2000. Nineteenth Annual Joint Conference of the IEEE Computer and Communications Societies. Proceedings. IEEE. Volume 3. (2000) 1107–1116 vol.3

17. Kohli, M., Lobo, J.: Realizing Network Control Policies Using Distributed Action Plans. Journal of Network and Systems Management **11** (2003)

18. Kanada, Y.: Policy Division and Fusion: Examples and a Method-or, Multiple Classifiers Considered Harmful. In: Integrated Network Management Proceedings, 2001 IEEE/IFIP International Symposium on. (2001) 545–560

19. Kanada, Y., O'Keefe, B.J.: Rule-Based Building-Block Architectures for Policy-Based Networking. Journal of Network and Systems Management **11** (2003)

20. Strassner, J.: Autonomic networking - theory and practice (tutorial seven). In: 2004 IEEE/IFIP Network Operations and Management Symposium. (2004)

21. Cuervo, F., Sim, M.: Policy Control Model: a Key Factor for the Success of Policy in Telecom Applications. In: IEEE 5th International Workshop on Policies for Distributed Systems and Networks (POLICY 2004). (2004)

22. Seagall, B., Arnold, D.: Elvin has left the building: A publish/subscribe notification service with quenching. In: Australian Unix and Open Systems Group. (1997)

23. Forgy, C.L.: Rete: A Fast Algorithm for the Many Pattern/Many Object Pattern Match Problem. Artificial Intelligence **19** (1982) 17–37

24. Miranker, D.P.: TREAT: A Better Match Algorithm for AI Production Systems. In: National Conference on Artificial Intelligence. Volume 1. (1987) 42–47

25. Perlin, M.: Incremental binding-space match: The linearized matchbox algorithm. In: Proceedings of the IEEE Conference on Tools for AI. (1991)

26. various: Design Principles for the Immune Systems and Other Distributed Autonomous Systems. Santa Fe Institute Studies in the Sciences of Complexity. Oxford University Press (2001)

27. Bonabeau, E., Dorigo, M., Theraulaz, G.: Swarm Intelligence: From Natural to Artificial Systems. Santa Fe Institute Studies in the Sciences of Complexity. Oxford University Press (1999)

Appendix

Most PBNM languages and the supporting inferencing engines do not provide default action semantics. That is, if the policy has no "default" clause that is invoked when the condition clause fails to evaluate to TRUE in response to an event. This may be appropriate for some applications, but many Telecoms policy applications involve what we term as a "Definite Response" policy pattern. That is, when presented with a particular question via an event, the policy server *must* present an answer ("permit/deny", "yes/no", "gold/silver/bronze" etc).

For such environments requiring to implement the SAC application, the policy described in 1. requires two rules to perform correctly:

```
policy R1 {
    condition {
        event Sac_Request: (srcAddr * * destPort protocol}
             UserIP      : (userName srcAddr}
             AppSpec     : (appName destPort protocol}
             Permit      : (userName appName}
    }
    action {
        Send(Sac_Response, SrcRouter, SessionID, Permit);
    }
}
policy R2 {
    condition {
        event Sac_Request: (srcAddr * * destPort protocol}
           not (UserIP    : (userName srcAddr}
               AppSpec    : (appName destPort protocol}
               Permit     : (userName appName}
               )
    }
    action {
        Send(Sac_Response, SrcRouter, SessionID, Deny);
    }
}
```

It is important that for *any* SacReq event presented to the policy server, only one of the two rules is activated and admitted into the Activation Set. An interesting question is how might one prove the correctness of the two rules under all conditions? We note the following pre and post conditions hold:

Pre-Condition: $|AS| = 0$ and $|SaqReq| = 1$
Post-Condition: $|AS| = 1$

Let

$$J = \sigma_{\substack{UserName \\ AppName}} \left(\sigma_{\substack{DestPort \\ Protocol}} \left(\sigma_{SrcAddr}(SacReq \times UserIP) \times AppSpec \right) \times Permit \right)$$

$$(12)$$

Then the operation of the two rules together may be described as:

$$AS = R1 \cup R2 \tag{13}$$

where

$$
\begin{aligned}
R1 &= J \\
R2 &= \sigma(SacReq \times \bar{J})
\end{aligned}
\tag{14}
$$

Therefore,

$$
\begin{aligned}
|AS| &= |J \cup \sigma(SacReq \times \bar{J})| \\
&\leq |J| + |\sigma(SacReq \times \bar{J})|
\end{aligned}
\tag{15}
$$

But the following observation holds:

− if $|J| = 1$ then $|\sigma(SacReq \times \bar{J})| = 0$, and
− if $|J| = 0$ then $|\sigma(SacReq \times \bar{J})| = 1$

Therefore $|AS| = 1$ under all conditions and the conditional operation of the two rules is shown to be correct.

Spatial Computing: An Emerging Paradigm for Autonomic Computing and Communication

Franco Zambonelli and Marco Mamei

DISMI - Università di Modena e Reggio Emilia,
Via Allegri 13, 42100 Reggio Emilia – Italy
{franco.zambonelli, mamei.marco}@unimore.it

Abstract. Emerging distributed computing scenarios call for novel "autonomic" approaches to distributed systems development and management. In this position paper we analyze the distinguishing characteristics of those scenarios, discuss the inadequacy of traditional paradigms, and elaborate on primary role of "space" in modern distributed computing. In particular, we show that spatial abstractions promise to be basic necessary ingredients for a novel "spatial computing" paradigm, acting as a unifying framework for autonomic computing and communication. On this base, we propose a preliminary "spatial computing stack" to frame the key concepts and mechanisms of spatial computing. Eventually, we try to sketch a research agenda in the area.

1 Introduction

In the past few years, a variety of novel distributed computing scenarios have emerged that, although apparently very different from each other, share some key characteristics. By considering scenarios as diverse as P2P networks, multi-agent ecologies, pervasive computing systems, sensor networks, and robot swarms, one can easily recognize that [18,20]: *(i)* they all involve distributed computational and communication activities taking place in decentralized networks with a very large number of components and *(ii)* with a highly dynamic structure; *(iii)* components are embedded in some external dynamic environment, whether physical or computational, and their activities are influenced by their position in that environment.

The large size and the dynamics of the network, as well as the unpredictability induced by environmental dynamics, make traditional approaches to distributed systems management – involving humans-in-the-loop and typically assuming the capability of centralized control – fall short. Novel approaches supporting autonomous self-configuration and self-adaptation of activities in response to network and environmental dynamics are required.

A variety of solutions exploiting specific forms of self-organization and self-adaptation to solve specific application problems are being proposed (see [3] for a comprehensive overview). The question of whether it is possible to devise a single unifying conceptual framework, applicable with little or no adaptations to scenarios as diverse as P2P networks and local networks of embedded sensors, is still open.

In this position paper, without having the ambition of providing a definitive answer to the above question, we will try to identify the important role that will likely

M. Smirnov (Ed.): WAC 2004, LNCS 3457, pp. 44–57, 2005.

be played in that process by spatial abstractions. In addition to the fact that spatial abstractions naturally suit systems whose activity are situated in some environment, the key point is that a spatial computing model can facilitate the integration of autonomic feature in distributed systems. In particular: *(i)* the central role of the network is substituted by an abstraction of space, built over the network in an autonomous and adaptive way; *(ii)* all application-level activities are abstracted as taking place in such space; *(iii)* autonomic behavior emerge form both the capability of the system of dynamically adapting the structure of the space as well as from the capability of application-level components of sensing, acting in, and navigating that space.

This paper is organized as follow. Section 2 outlines the key characteristics of modern scenarios and discusses the inadequacy of traditional approaches. Section 3 introduces "Spatial Computing" and discusses its basic concepts and advantages, and its relations with autonomic computing and communications. Section 4 proposes a "Spatial Computing Stack", as a framework to organize and understand the basic abstractions and mechanisms involved in spatial computing. Section 5 sketches a rough research agenda in the area and concludes.

2 Modern Distributed Systems Scenarios and the Need for Novel Approaches

A variety of modern distributed computing scenarios exhibit characteristics challenging traditional approaches to network and distributed systems management.

2.1 Key Characteristics

Such scenarios include *(i)* micro-scale ones, i.e., networks of low-end computing devices typically distributed over a geographically small area (e.g., sensor networks [4], smart dusts [10] and spray computers [17, 18]); *(ii)* medium-scale scenarios, i.e., networks of medium-end devices, distributed over a geographically bounded area, and typically interacting with each other via short/medium range wireless connections (pervasive computing systems and smart environments [5] and cooperative robot teams); *(iii)* global-scale scenarios, characterized by high-end computing systems interacting at a world-wide scale (the physical Internet, the Web, P2P networks [13] and multiagent systems ecologies [6].

Despite clear dissimilarities in structure and goals, one can also easily recognize some key common characteristics:

- **Large Scale:** the number of nodes in all the above types of networks and consequently the number of components involved in a distributed application is typically very high and, due to decentralization, hardly controllable. It is not possible neither to enforce a strict control over the configuration of components (consider e.g., the nodes of a P2P network) nor to directly control each of them during execution (consider e.g., the nodes of a sensor network distributed in a landscape).
- **Network dynamism:** the activities of components will take place in network whose structure derives from an almost random deployment process, and that is

likely to change over time with unpredictable dynamics. Factors that may contribute to dynamically change the network topology include environmental contingencies or failure of components (very likely e.g., in sensor networks and pervasive computing systems) and mobility of nodes (as e.g., in robot teams and in network of smart appliances). In addition, at the application level, software components can be of an ephemeral or temporary nature (consider e.g. the peers of a P2P network).

- **Situatedness:** The activities of components will be strongly related to their location in either a physical or a virtual computational environment. On the one hand, situatedness can be at the very core of the application (e.g. in sensor networks and in pervasive computing systems the very goal is to exploit the physical location of nodes and their capabilities to collect environmental data and to improve our interaction with the physical world). On the other hand, situatedness can relate to the fact that components can take advantage of their environment to organize the access to distributed resources (as e.g., in P2P data sharing networks).

The first two characteristics (large size and network dynamism) compulsory call for self-organizing and self-adapting approaches, enabling those systems to exhibit – both at the network and at the application level – autonomic behavior. In fact, if the dynamics of the network and of the environment compulsory require dynamic adaptation, the impossibility of enforcing a direct control over each component of the system implies that such adaptation must occur without any human intervention. The last characteristic, situatedness, calls for an approach that elects the environment, its spatial distribution, and its dynamics, to primary design dimensions, aspects which have been mostly disregarded by traditional approaches. In any case, the capability of self-organization and self-adaptation cannot abstract from the capability of the system of becoming "context-aware", i.e., of letting components to perceive the local properties of the space in which they are situated, and to act and adapt their behavior on this basis.

Summarizing: all presented scenarios of distributed computing share very similar characteristics and all require novel approaches promoting both autonomic behavior and an explicit modeling of situatedness. On this base, one could imagine that a single general-purpose distributed computing and communication paradigm, suitable for a variety of scenarios and enabling to face a variety of problems in a uniform way can be conceived. Unfortunately, traditional distributed computing paradigm appears not suitable to this purpose.

2.2 Inadequacy of Traditional Approaches

Early researches in parallel and distributed computing promoted a *transparent distributed computing* paradigm, in which the presence of an underlying network was totally hidden from application components [2]. The key motivation was that, to avoid the complexities inherent in having to deal with a distributed environment, it was necessary to hide distribution and enable components to execute and interact with each other as if they were all executing on a single, centralized, node. Figure 1a summarizes this by outlining that a component can interact with another one by simply "naming" it and disregarding its actual position.

Unfortunately, a transparent approach to distributed computing is totally unsuitable for modern scenarios. First, promoting transparency is very costly and can hardly scale to large-scale systems, in that it requires the presence of complex global naming services. In addition, transparent – i.e., "by name" – interactions can be effectively supported only in the presence of static interaction patterns and closed systems, definitely not in the presence of network dynamics. Finally, a transparent distributed computing paradigm does not provide an appropriate abstraction to deal with the situatedness of components in a distributed environment: under this paradigm, the distributed environment does not exist.

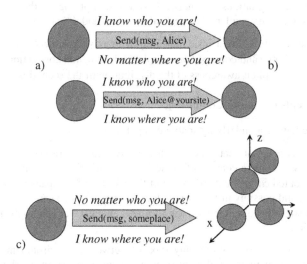

Fig. 1. Interactions in: (a) a transparent distributed computing model; (b) a network-aware distributed computing model; (c) a spatial computing model

The limitations of transparent distributed computing became evident in the mid 90's, with the advent of the Internet and of world-wide distributed computing. Such global scenarios outlined the need for *network-aware computing* models, in which application components were made aware of the distributed and decentralized nature of their operational environment [16]. The assumption in network-aware computing is to make all interactions rely on the explicit knowledge of the network allocation – i.e., the IP – of components and resources. Figure 1b summarizes this with regard to the interaction between two distributed components: an interaction relies on the knowledge on the local name of a component on a specific network node.

Network-aware computing is suitable for large-scale network systems. First, it enables to explicitly take into account the costs involved in distributed interactions. Second, it involves local naming services which can scales well. However, network-aware computing does not provide at hand solutions to deal with network dynamics. If the structure of the network can dynamically change, and if nodes (and software components over them) can be of an ephemeral nature (i.e., intermittently available) a network-aware computing model requires to handle explicitly the exceptions caused by nodes unavailability. This naturally complicates the system's execution and management and also increases the costs and complexity of discovery services. Also, as far as situatedness is concerned, the only actual environmental abstraction reduces to be the network itself, an abstraction which is quite low level and does not easily enable modeling the logical relations between the network components.

3 Spatial Computing

To overcome the limitations of network-aware computing without losing its advantages it is necessary to identify a general-purpose model that:

- Hides the complexities intrinsic in dealing with a dynamic network.
- Provides suitable and conceptually simple environmental abstractions.
- Preserves an explicit awareness of the distribution of the scenarios.

3.1 Key Concepts in Spatial Computing

To this end, the key idea underlying spatial computing is to:

- Make the concept of "network" – a discrete system of variously interconnected nodes – evolve into a concept of "space" – i.e., a metric continuum.
- Let distributed components be aware of their surrounding space and to let them perceive (and possibly influence) the local properties of space.
- Rely on such spatial perception for all management-level and application-level activities.

In particular, in spatial computing, any type of networked environment is hidden below some of virtual metric n-dimensional space, mapped as an overlay over the physical network. The nodes of the network are assigned a specific area of the virtual space, and are logically connected to each other accordingly to the spatial neighborhood relations. Accordingly, each and every entity in the network, being allocated in some nodes of the network, is also automatically situated in a specific position in space. In this way, components in the network are no longer "network-aware" but rather "space-aware". On the one hand, components perceive their local position in space as well as the local properties of space (e.g., the locally available data and services) and possibly change them. On the other hand, the activities of components in that space are related to some sort of "navigation" in that space, which may include moving themselves to a specific different position of space or moving data and events in space according to "geographical" routing algorithms. The primary way to refer to entities in the network is by "position", i.e., any entity is characterized by being situated in a specific position in the physical space. In other words, the concept of "names" loses its primary role. This is summarized in Figure 1c: an entity interacts to another entity by sending data to a position in space.

Spatial computing models appear very suitable for the identified key characteristics of modern distributed computing scenarios:

- **Large size:** the size of a network does not influence the models or the mechanisms, which are the same for a small network and for a dramatically large one.
- **Network dynamics:** since the presence of the network is not directly perceived by components, the fact that it can be of a highly dynamic nature is irrelevant. The network is hidden behind a stable structure of space that is maintained despite network dynamism.
- **Situatedness:** the abstraction of space is a conceptually simple abstraction of environment, which also perfectly matches the needs of those systems, such as pervasive computing systems and sensor networks, whose activities are strictly intertwined with the physical space.

3.2 Examples of Spatial Computing Approaches

We do not claim to have invented the spatial computing paradigm from scratch. Rather, we consider spatial computing as an emerging trend that is, more or less explicitly, making its appearance in diverse scenarios.

As an example, consider a sensor network scenario with a multitude of wireless sensors randomly deployed in a landscape to perform some monitoring of environmental conditions [4]. There, all activities of sensors are intrinsically of a spatial nature. First, each sensor is devoted to local monitoring a specific portion of the physical space (that it can reach with its sensing capabilities). Second, components must coordinate with each other based on their local positions, rather than on their IDs, to perform activities such as detecting the presence and the size of pollution clouds, and the speed of their spreading in the landscape. All of this implies that components must be made aware of their relative positions in the spatial environment by re-constructing a virtual representation of the physical space [9]. Moreover, they can take advantage of "geographical" communication and routing protocols in which messages, data, and events, flow towards specific position of the physical/virtual space [11].

Another example in which spatial concepts appear in a less trivial way is world-wide P2P computing. In P2P computing, an overlay network of peers is built over the physical network and, in that networks, peers act cooperatively to search specific data and services. In first generation P2P systems (e.g., Gnutella [13]), the overlay network is totally unstructured, being built by having peers randomly connect to a limited number of other peers. Therefore, in these networks, the only effective way to search for information is message flooding. More recent proposals [12, 14] suggest structuring the network of acquaintances into specific regular "spatial shapes", e.g., a ring or an N-dimensional torus. When a peer connects to the networks, it occupies a portion of that spatial space, and networks with those other peers that are neighbors accordingly to the occupied position of space. Then, data and services are allocated in specific positions in the network (i.e., by those peers occupying that position) depending on their content/description (as can be provided by a function hashing the content into specific coordinates). In this way, by knowing the shape of the network and the content/description of what data/services one is looking for, it is possible to effectively navigate in the network to reach the required data/services. That is, P2P networks define a spatial computing scenario in which all activities of application components are strongly related to positioning themselves and navigating in an abstract metric space. It is also worth outlining that recent researches promote mapping such spatial abstractions over the physical Internet network so as to reflect the geographical distribution of Internet nodes (i.e., by mapping IP addressed into geographical physical coordinates [15]) and, therefore improve efficiency.

In addition to the above examples, other proposals in areas such as pervasive computing [1] and self-assembly [9] explicitly exploit spatial abstractions (and, therefore, a sort of spatial computing model) to organize distributed activities.

3.3 Autonomic Features in Spatial Computing

Autonomic features, including the capability of a distributed system of self-configuring its activity, self-inspecting and self-tuning its behavior in response to changed conditions, or self-healing it in the presence of faults, are necessary for enabling spatial computing and, at the same time, are also intrinsically promoted by the adoption of a spatial computing model.

On the one hand, to enable a spatial computing model, it is necessary to envision mechanisms to build the appropriate overlay spatial abstraction and to have such spatial abstraction be coherently preserved despite network dynamics. In other words, this requires the nodes of a network to be able to autonomously connect with each other, set up some sort of common coordinate systems, and self-position themselves in such space. In addition, this requires the nodes of the network to be able to self-reorganize their distribution in the virtual space so as to *(i)* make room for new nodes joining the network (i.e., allocate a portion of the virtual space to these nodes); *(ii)* fill the space left by nodes that for any reason leave the network; *(iii)* re-allocate the spatial distribution of nodes to react to node mobility. It is also worth outlining that, since the defined spatial structure completely shields the application from the network, it is also possible for a system to dynamically tune the structure of the space so as enforce some sorts of autonomic management of the network, transparently to the higher application levels. As an example, load unbalances in the network can be dynamically dealt, transparently from the application level, by simply re-organizing the spatial structure so as to have overloaded nodes occupy a more limited portion of the space.

On the other hand, the so defined spatial structure can be exploited by application level components to organize their activities in space in an autonomous and adaptive way. First of all, it is a rather assessed fact that "context-awareness" and "contextual activity", i.e., the capabilities of a component to perceive the properties of the operational environment and of influencing them, respectively, are basic ingredients to enable any form of adaptive self-organization and to establish the necessary feedback promoting self-adaptation. In spatial computing, this simply translates in the capability of perceiving the local properties of space, which in the end reflect some specific characteristics of either the network or of some application-level characteristics and of changing them. Second, one should also recognize that the vast majority of known phenomena of self-organization and self-adaptation in nature (from ant-foraging to reaction-diffusion systems, just to mention two examples in biology and physics) are actually phenomena of self-organization in space, emerging from the related effect of some "component" reacting to some property of space and, by this reaction, influencing at its turn the properties of space. Clearly, a spatial computing model makes it rather trivial to reproduce in computational terms such types of self-organization phenomena, whenever they may be of some use in a distributed system.

4   Framing Spatial Computing

Let us now have a detailed look at the basic mechanisms that have been exploited so far in distributed computing to promote self-organization and autonomic behavior. We will show that most of these mechanisms can be easily interpreted and mapped into very similar spatial concepts, and that they can be framed in a unifying flexible framework.

4.1   A Spatial Computing Stack

As shown in Figure 2, we frame these mechanisms according to a "space-oriented" stack of levels. For each level, we can recognize that different mechanisms, exploited in different scenarios, can serve similar purposes and aim at providing similar "spatial services" (see Table 1). In other words, by introducing a new paradigm rooted on spatial concepts, it is possible to interpret a lot of proposed self-organizing and autonomic approaches in terms of mechanisms to manage and exploit the space. On this basis, it is likely that a simply unifying model for autonomic computing and communication – leading to a single programming model and methodology and – can be actually identified.

The lowest *"physical level"* is about how components start interacting – in a dynamic and spontaneous way – with other components in the systems. This is a very basic expression of autonomic behavior which is a pre-requisite to support more complex forms of autonomy and of self-organization at higher levels. To this end, the basic mechanism exploited is broadcast (i.e. communicate with whoever is available). Radio broadcast is used in sensor networks and in pervasive computing systems, and different forms of TCP/IP broadcast (or of dynamic lookup) are used as a basis for the establishment of overlay networks in wide area P2P computing. Whatever the case, this physical level can be considered as in charge of enabling a component of a dynamic network application to get into existence in it and to start interacting with the other components.

The *"structure level"* is the level at which some sort of spatial structure is built and maintained by components existing in the physical network. As already outlined, the fact that a system is able to create a stable spatial structure capable of surviving network dynamics and adapting the working conditions of the network is an important expression of autonomic behavior *per se*. However, such spatial structure is not a goal for the application, and it is instead used as the basic spatial arena to support higher levels of organization of activities.

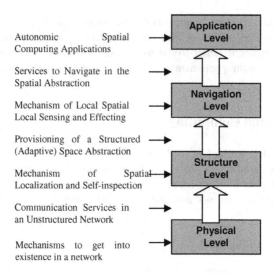

Autonomic   Spatial   →
Computing Applications

Services to Navigate in the   →
Spatial Abstraction

Mechanism of Local Spatial   →
Local Sensing and Effecting

Provisioning of a Structured   →
(Adaptive) Space Abstraction

Mechanism   of   Spatial   →
Localization and Self-inspection

Communication Services in   →
an Unstructured Network

Mechanisms to get into   →
existence in a network

Fig. 2. A Spatial Computing Stack

The various mechanisms that are used at the structure level in different scenarios are – again – very similar to each other. Sensor networks as well as self-assembly systems typically structure the space accordingly to their positions in the physical space, by exploiting mechanisms of geographical self-localization. Pervasive computing systems, in addition to mechanisms of geographical localization, often exploit logical spatial structures reflecting some sorts of abstract spatial relationships of the physical world (e.g., rooms in a building) [1]. Global scale systems, as already anticipated, exploits overlay networks built over a physical communication network. Although early approaches (e.g., Gnutella) give no metric structure to such overlay space, more recent approaches (as already anticipated) typically exploit metric overlay spaces and aims at making the spatial overlay match the spatial distribution of Internet nodes. A possibility which is currently under-investigated relates to the possibility – at the structure level – of dynamically adapting the structure of the space (and not simply of preserving a stable structure) to reflect and adapt to changing working conditions in the network.

The "*navigation level*" concerns the basic mechanisms that components exploit to orient their activities in the spatial structure and to sense and affect the local properties of space (i.e. mechanism to actually "use" the available spatial structure). If the spatial structure has not any well-defined metric, the only navigation approaches are flooding and gossiping. However, if some sort of metric structure is defined at the structure level (as, e.g., in the geographical spatial structures of sensor networks or in metric overlay networks) navigation approaches typically relate in following the metrics defined at the structure level. For instance, navigation can imply the capability of components to reach specific points (or of directing messages and data) in the space based on simple geometric considerations as in, e.g., geographical routing.

Table 1. Spatial Mechanisms in Modern Distributed Computing Scenarios

	MICRO SCALE Nano Networks, Sensor Networks, Smart Dust, Self-Assembly, Modular Robots	MEDIUM SCALE Home Networks, MANETs, Pervasive Environments, Mobile Robotics	GLOBAL SCALE Internet, Web, P2P networks, multiagent systems
"Application" Level (exploiting the spatial organization to achieve in a *self-organizing and adaptive way* specific app. goals)	Spatial Queries Spatial Self-Organization and Differentiation of Activities Spatial Displacement Motion Coordination & pattern formation DATA: environmental data	Discovery of Services Spatial Displacement Coordination and Distribution of Task and Activities Motion coordination & pattern formation DATA: local resources and environmental data	P2P Queries as Spatial Queries in the Overlay Motion Coordination on the Overlay Pattern formation (e.g., for network monitoring) DATA: files, services, knowledge
"Navigation" Level (dealing with the mechanism exploited by the entities living in the space to *direct activities and movements in that space*)	Flooding Gossiping (random navigation) Geographical Routing (selecting and reaching specific physical coordinates) Directed Diffusion (navigation following sorts of computational fields) Stigmergy (navigation following pheromone gradients)	Computational fields Multi-hop routing based on Spanning Trees Pattern-matching and Localized Tuple-based systems	Flooding Gossiping (random navigation) Metric-based (moving towards specific coordinates in the abstract space) Gossiping (random navigation) Stigmergy (navigation following pheromone gradients distributed in the overlay network)
"Structure" Level (dealing with mechanisms and policies to adaptively *shape a metric space* and let components find their position in that space)	Self-localization (beacon-based triangulation)	Self-localization (Wi-Fi or RFID triangulation) Definition and Maintenance of a Spanning Tree (as a sort of navigable overlay)	Establishment and Maintenance of an Overlay Network (for P2P systems) Referral Networks and e-Institutions (for multiagent systems)
"Physical" Level (dealing with the mechanism necessary to *get into existence* in a network)	Radio Broadcast Radar-like localization	Radio Broadcast RF-ID identification	TCP broadcast – IP identification Directed TCP/UDP messages Location-dependent Directory services

Starting from the basic navigation capability, is also possible to enrich the structure of the space by propagating additional information to describe "something" which is happening in that space, and to differentiate the properties of the space in different areas. One can say that the structure of space may be characterized by additional types of spatial structures propagating in it, and that components may direct their activities based on navigating these additional structures. In other words, the basic navigation capabilities can be used to build additional spatial structures with different navigation mechanisms. Typical mechanisms exploited at these additional levels are computational fields and pheromones. Despite the different inspiration of the two approaches (physical versus biological), we emphasize that they can be modeled in a uniform way, e.g., in terms of time-varying properties defined over a space [7]. The basic expression of self-organization that arises here derives from the fact that the structures propagated in the space – and thus the navigation activity of application components – are updated and maintained to continuously reflect the actual structure and situation of the space.

At the *"application level"*, navigation mechanisms are exploited by application components to interact and organize their activities. Applications can be conveniently built on the following self-organizing feedback loop: *(i)* having components navigate in the space (i.e., discriminating their activities depending on the locally perceived structure and properties of the space) and *(ii)* having components, at the same time, modifying existing structure due to the evolution of their activities.

Depending on the types of structures propagated in the space, and on the way components react to them, different phenomena of self-organization can be achieved and modeled. For example, processes of morphogenesis (as needed in self-assembly, modular robots and mobile robotics), phenomena mimicking the behavior of ant-colonies and of flocks, phenomena mimicking the behavior of granular media and of weakly correlated particles, as well as a variety of social phenomena, can all be modeled in terms of:

- entities getting to existence in a space;
- having a position in a structured space and possibly influencing its structure;
- capable of perceiving properties spread in that space;
- capable of directing their actions based on perceived properties of such space and capable of acting in that space by influencing its properties at their turn.

Still, the ultimate goal of a uniform modeling approach capable of effectively capturing the basic properties of self-organizing computing, and possibly leading to practical and useful general-purpose modeling and programming tools, is far from close.

4.2 Multiple Spaces and Nested Spaces

In general, different scenarios and different application problems may require different perceptions of space and different spatial structures. For instance, a world-wide resource-sharing P2P network over the Internet may require – for efficiency reason – a 2-D spatial abstraction capable of reflecting the geographical distribution of Internet nodes over the earth surface. On the other hand, a P2P network for social interactions may require a spatial abstraction capable of aggregating in close regions of the virtual space users with similar interests. Also, one must consider that in the near future, the different network scenarios we have identified will be possibly part of a unique huge network (consider that IPv6 addressing will make it possible to assign an IP address to each and every square millimeter on the earth surface). Therefore, it is hard to imagine that a unique flat spatial abstraction can be effectively built over such a network and satisfy all possible management and application needs.

With this regard, the adoption of the spatial computing paradigm does not prescribe at all to adopt the same set of mechanisms and the same type of spatial structure for all networks and for applications. Instead, being the spatial structure a virtual one, it is possible to conceive both *(i)* the existence, over the same physical network, of multiple complimentary spatial abstraction independently used by different types of applications; and *(ii)* the existence of multiple layers of spatial abstractions, built one over the other in a multi-layered system.

With regard to the former point, in addition to the example of the different types of P2P networks calling for different types of spatial abstractions, one could also think at how different problems such as Internet routing, Web caching, virtual meeting points, introduce very different problems and may require the exploitation of very different spatial concepts.

With regard to the latter point, one can consider two different possibilities. Firstly, one can think at exploiting a first-level spatial abstractions (and the services it provides) to offer a second-level spatial abstraction enriching it with additional specific characteristics. For examples, one can consider that a spatial abstraction capable of mapping the nodes of the Internet into geographical coordinates can be exploited, within a campus, to build an additional overlay spatial abstraction mapping such coordinates into logical location (e.g., the library, the canteen, the Computer Science department and, within it, the office of Prof. Zambonelli). Such additional spatial abstraction could then be used to build semantically-enriched location dependent services. Secondly, one could think at conceiving a hierarchy of spatial abstractions that provides different levels of information about the space depending on the level at which they are observed, the same as the information we get on a geographical region are very different depending on the scaling of the map on which we study it. As an example, we can consider that the spatial abstraction of a wide-area network can map a sensor network – connected to the large network via a gateway – as a "point" in that space, and that the distributed nature of the sensor networks (with nodes having in turn a specific physical location in space) becomes apparent only when some activity takes place in that point of space (or very close to it).

In any case, although we have strongly advocated the flexibility and modularity of a spatial computing paradigm, whether and how it could be put to practice and could fulfill its promises is an open research issue.

5 Research Agenda

Spatial computing, by abstracting the execution of distributed applications around spatial concepts (localization and navigation in some sorts of metric space), and by exploiting the same set of spatial abstractions to perform adaptive network management activities, promises to be an effective paradigm for modern distributed computing scenarios.

In any case, besides the considerations made in this paper, much formal and practical work is needed to asses the potentials of spatial abstractions in distributed computing, and to verify whether they can actually pave the way to a sound and general-purpose engineered approach to autonomic computing and communication. In particular:

- Is the research for a unifying model fueled by enough application problems? In other words, is there a compulsory need for a unifying approach to promote a uniform autonomic treatment of a variety of problems in different scenarios? Or is instead the current way of doing (i.e., researching specific special-purpose autonomic solutions to specific problems in specific scenarios) the most economic and effective one?

- If the research of a unifying modes is worth (as we believe), is the proposed spatial computing stack meaningful and useful? Here, we have proposed it as a preliminary attempt to frame some basic concepts, and we are well aware that not everything fits perfectly in it. Nevertheless, our opinion is that, once all the layers will be properly defined, they will support a better engineering of such systems, promoting separation of concerns and clearly identifying the duties of the different levels.

- Beside the fact that spatial computing seems promising, can really all (or at least a large portion of) conceivable autonomic features – from the network management level up to the application level – be effectively expressed and implemented in such terms? With regard to network management, earlier in this paper we have sketched a rough idea on how spatial computing could be of some use to adaptively promote load balancing in a network. However, could issues such as QoS management, service personalization, emergence of pathological congestion patterns, take any advantage of a spatial computing approach? With regard to the application level, a variety of application problems requiring self-organizing and adaptive behavior deals with concepts that can be hardly intuitively mapped into spatial concepts. Consider, for instance, phenomena of adaptive division of labor or phenomena or adaptive evolution. For these problems, would exploring some sorts of spatial mapping to face them still be useful and practical? Would it carry advantages?

- If a unifying model spatial computing model can be found, can it be translated into a limited set of general-purpose and manageable protocols, tools, and programming abstractions? Also, can it lead to the identification of a sound methodology for developing and deploying autonomic features in modern distributed systems?

All the above issues need to be investigated, complemented by researches aimed at better understanding and framing the behavior of self-organizing and self-adaptive systems.

To conclude, we simply point out that our research group is currently involved in the development of a middleware called TOTA [8]. TOTA, suitable for both large-scale wide-area networks and local ad-hoc networks of small computing devices, defines *(i)* a set of network-level services to build and self-maintain spatial overlay structures over dynamic networks; *(ii)* an API to be exploited by application agents to sense and effect the local properties of space and to navigate in the spatial structure. In particular, TOTA exploits sorts of potential fields to be propagated in space to implement in a simple and modular way a variety of overlay spatial abstractions, and it has been successfully experienced to achieve self-organization and self-adaptation in a variety of distributed applications.

Acknowledgments. Work Supported by the Italian MIUR in the context of the "Progetto Strategico ICT: IS-MANET: Infrastructures for Mobile Ad-Hoc Networks".

References

1. C. Borcea, "Spatial Programming Using Smart Messages: Design and Implementation", *24th Int.l Conference on Distributed Computing Systems*, Tokio (J), May 2004.
2. R. S. Chin, S. T. Chanson, "Distributed Object-Based Programming Systems", *ACM Computing Surveys*, 23(1), March 1991.
3. G. Di Marzo, A. Karageorgos, O. Rana, F. Zambonelli (Eds.), *Engineering Self-organizing Systems: Nature Inspired Approaches to Software Engineering*, LNCS No. 2977, Springer Verlag, May 2004.
4. D. Estrin, D. Culler, K. Pister, G. Sukjatme, "Connecting the Physical World with Pervasive Networks", *IEEE Pervasive Computing*, 1(1):59-69, 2002.
5. H.W. Gellersen, A. Schmidt, M. Beigl, "Multi-Sensor Context-Awareness in Mobile Devices and Smart Artefacts", *Mobile Networks and Applications*, 7(5): 341-351, Oct. 2002.
6. J. Kephart, "Software Agents and the Route to the Information Economy", *Proceedings of the National Academy of Science*, 99(3):7207-7213, May 2002.
7. M. Mamei, F. Zambonelli, "Co-Fields: a Unifying Approach to Swarm Intelligence", *3rd Workshop on Engineering Societies in the Agents' Word*, LNCS No. 2677, Springer Verlag, April 2003.
8. M. Mamei, F. Zambonelli, "Programming Pervasive and Mobile Computing Applications with the TOTA Middleware", *2nd IEEE Conference on Pervasive Computing and Communications*, Orlando (FL), IEEE CS Press, March 2004.
9. R. Nagpal, H. Shrobe, J. Bachrach, "Organizing a Global Coordinate System from Local Information on an Ad Hoc Sensor Network", *2nd International Workshop on Information Processing in Sensor Networks*, Palo Alto (CA), April, 2003.
10. K. Pister, "On the Limits and Applicability of MEMS Technology", *Defense Science Study Group Report*, Institute for Defense Analysis, Alexandria (VA), 2000.
11. A. Rao, C. Papadimitriou, S. Ratnasamy, S. Shenker, I. Stoica. "Geographic Routing Without Location Information". *ACM Mobicom Conference*. San Diego (CA), USA, 2003.
12. S. Ratsanamy,, P. Francis, M. Handley, R. Karp, "A Scalable Content-Addressable Network", *ACM SIGCOMM Conference 2001*, Aug. 2001.
13. M. Ripeani, A. Iamnitchi, I. Foster, "Mapping the Gnutella Network", *IEEE Internet Computing*, 6(1):50-57, Jan.-Feb. 2002.
14. A. Rowstron, P. Druschel, "Pastry: Scalable, Decentralized Object Location and Routing for Large-Scale Peer-to-Peer Systems", *18th IFIP/ACM Conference on Distributed Systems Platforms*, Heidelberg (D), Nov. 2001.
15. A. Rowstron et al., "PIC: Practical Internet Coordinates", *24th International Conference on Distributed Computing Systems*, IEEE CS Press, Tokyo (J), May 2004.
16. J. Waldo et al., "A Note on Distributed Computing", *Mobile Object Systems*, LNCS No. 1222, Feb. 1997.
17. F. Zambonelli, M.P. Gleizes, R. Tolksdorf, M. Mamei, "Spray Computers: Explorations in Self-organization", *Journal of Pervasive and Mobile Computing*, 1(1), May 2005.
18. F. Zambonelli, N. Jennings, M. Wooldridge, "Developing Multiagent Systems: the Gaia Methodology", *ACM Transactions on Software Engineering and Methodology*, 12(3):410-470, July 2003.
19. F. Zambonelli, M. Mamei, "The Cloak of Invisibility: Challenges and Applications", *IEEE Pervasive Computing*, 1(4):62-70, Oct.-Dec. 2002.
20. F. Zambonelli, V. Parunak, "Towards a Paradigm Change in Computer Science and Software Engineering", *The Knowledge Engineering Review*, 18(4):329-342, 2004.

Self-deployment, Self-configuration: Critical Future Paradigms for Wireless Access Networks

Francis J. Mullany, Lester T.W. Ho,
Louis G. Samuel, and Holger Claussen[1]

Bell Labs Research, Lucent Technologies,
The Quadrant, Stonehill Green, Westlea, Swindon,
SN5 7DJ, United Kingdom
{mullany, lho1, lsamuel, claussen}@lucent.com

Abstract. To combat the increasing significance of deployment and configuration costs, the concept of a self-deploying, self-configuring radio access network is discussed. It is proposed that the basic sciences of complex systems (cellular automata, game theory, ecology modeling) can be exploited to design algorithms for such a system. An example, taken from the field of cellular automata, is presented for a network capable of self-adaptation to achieve universal radio coverage in a simplified environment.

Keywords: Radio access networks; auto-configuration; self-deployment; self-organization; cognizant networks; complexity theory; game theory; cellular automata; ecology modeling.

1 Introduction

A number of trends for wireless access networks are clearly emerging: (a) reductions in equipment costs, (b) reduced cell sizes (with a commensurate increase in the total number of cells), and (c) additional complexity as interoperability between heterogeneous systems becomes economically critical. These all will increase the relative cost of deployment and configuration of the radio access nodes (base stations, access points, etc.), perhaps to the point where additional innovations will be difficult to introduce. Self-configuring radio access nodes help, but there is a strong need for the additional, novel concept of a *self-deploying* network. Such a network would, from experience of past traffic, be able to decide autonomously the changes needed in both the location and configuration of its wireless access nodes, and suggest locations for new nodes. Thus, this would be an innovative self-aware network that designs its own layout and configuration, adapting and expanding according to changes in user demand.

This vision differs from traditional ad-hoc networking concepts in that ad-hoc networks seek the optimal configuration, given the current location of nodes. Here the access network is allowed more freedom: the freedom to choose the location and

---

[1] Author for all correspondence. Tel: +44-1793-776406. Fax: +44-1793-776725.

M. Smirnov (Ed.): WAC 2004, LNCS 3457, pp. 58–68, 2005.

nature of the nodes needed. This also contrasts even more with the current state-of-the-art where either such planning is done (e.g. in cellular networks) in a quasi-manual manner with a mixture of off-line planning tools, expensive drive-testing and economic rules-of-thumb; or is done not at all (e.g. in WLANs – wireless local area networks) and low efficiency results. To move beyond this, the main algorithmic challenges are with respect to the expensive resources such as spectrum and back-haul capability. The algorithms used must be simple, distributed, robust to changing user demand and to heterogeneity in the underlying technology, and, above all, financially viable in a multi-operator environment.

The key theoretical frameworks that should be exploited to design such algorithms are complexity theory, small-world theory, cellular automata, game theory, microeconomic modeling, and ecology/population growth modeling. Some of these frameworks (e.g. game theory) will give insights into the algorithms to be used at the individual access nodes, while others (e.g. ecology modeling) will be critical in assessing the technical robustness and the financial viability of the resulting solutions. It is only through proving the robustness and economic viability that one can have confidence that the resulting network be able to drive its own design in a manner that meets the needs of its owners.

In this paper, the expected future developments in next-generation networks and the resulting drivers for self-deploying and self-configuring networks are explored in more detail in section 2. Section 3 discusses the various applicable basic sciences that may be used to achieve such solutions. A useful case study involving the application of concepts from cellular automata for achieving coverage in a simplified propagation environment is given in section 4. Conclusions are drawn in section 5.

2   Problem Statement

Figure 1 shown below is indicative of many of today's wireless access networks: These are vertically integrated to provide tightly controlled services like voice and as such are (a) extremely centralized in terms of control, (b) hierarchical in terms of architecture, (c) isolated with respect to other systems, and (d) very inflexible as far as adapting to new services and traffic demands.

However, it is clear that mobile communication systems will become richer in features and capability, and the isolation between systems will have to decrease [1]. Within systems, the need for rapidly deployable systems in areas of high-traffic density has pushed architectures towards flatter designs (e.g. 802.11b WLANs). Economic necessity will force operators and service providers to use more flexibility in their systems so as to keep up with changes in user needs and terminal capabilities. Figure 2 below shows the potential future vision. It is also significant that to increase capacity, average cell sizes are decreasing: witness the shrinking cell sizes of cellular systems in going from second- to third-generation technology and the introduction of WLANs and PANs (personal area networks).

All of the above inevitably will lead to exploding complexity in the management, construction, and configuration of these networks. Manual decision making and optimization will prove to be exorbitantly expensive and will end up dominating the total network costs, particularly as the capital expenses are reduced over time through improvements in hardware and software technologies. More fundamentally, the increases in complexity may exceed the capabilities of manual planning and configuration entirely, resulting in reductions in reliability and end-user trust of the system.

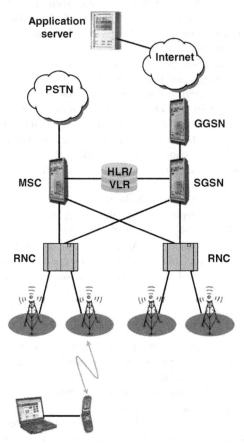

Fig. 1. Typical hierarchical, centralized, inflexible cellular network of today

Thus, if future wireless communications networks are to be viable financially and are to command the confidence of end users then the expected ad-hoc, dynamic architectures need to be highly robust, self-deploying and self-healing, with nodes that are auto-configurable and flexible. A *self-deploying* radio access network is one that is able to learn from its current performance, both technically (in terms of coverage and capacity) and economically (i.e. is the network profitable?) and then is able to determine what changes, additions, and removals of access nodes are needed[2] as user demands and the competitive environment changes over the long term, say weeks to years. *Self-configuration* is more of a short-term activity over tens of minutes to days: A node dropped into a coverage area must be able to integrate itself into an existing network quickly and reliably. The removal of a node from the network (e.g. through node failure) should also trigger a sequence of auto-configuration among the remaining nodes. Fundamentally, the nodes need to work together to adapt the instantaneous network configuration to short-term variations in the user traffic, readjusting to optimize the radio coverage, the traffic-bearing capacity, and also connectivity among the nodes in the network. The result should be a network that inspires end-user trust and confidence in its ability to provide, on demand, network transport over a wide range of conditions.

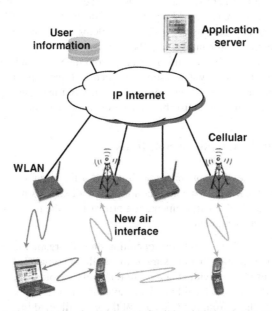

Fig. 2. Future vision of a distributed, flexible wireless access network

These objectives are critical enablers of economical deployment of complex communications systems, particularly in the following areas:

- Without the automation implied in the above vision, operators and wireless service providers will be very reluctant to exploit the potential advantages of heterogeneous air interface access and dynamic, demand-adaptive network architectures if the deployment costs and running costs explode as a result of the associated complexity.

---

[2] While this network can determine what changes are needed, it is expected that it will still generally require human intervention to implement the actual physical relocations! However, exceptions to this may occur in the fields of military and emergency communications.

- The situation is even more critical for small- to medium-sized businesses and non-profit organizations. The technology proposed here will allow them to overcome their inability to afford the highly specialized staff needed for manual deployment and configuration. Hence, they will be able to more easily deploy and exploit state-of-the-art internal wireless communication infrastructures within their premises and beyond.
- Automatic deployment and recovery is of extreme importance for flexible, quickly deployable emergency communication systems – a recognized key component for modern health and civil defense services.

At one level, these are problems of network architectures, interfaces, network protocols, and software and hardware architectures. There already exist substantial research efforts in these areas, e.g. [2], particularly in the field of self-configuration. However, there are also serious algorithmic problems in a few key areas: namely those where there are significant resource bottlenecks: e.g. air interface capacity and coverage, back-haul capabilities. Fundamentally, both the wired and wireless links of access networks are both costly and resource-limited and matching these resources to dynamic user demand is a non-trivial problem – on any timescale. This is exacerbated by the trend towards more distributed, heterogeneous networks.

3   Synthesis and Analysis of Solutions

The sciences of complex systems have much potential for providing solutions to the above challenges of self-deploying and -configuring radio access networks. In terms of synthesizing algorithms for implementation primarily at the node level, the following areas are promising candidates:

- *Cellular automata (CA).* A network operating on self-organizing behavior would have many characteristics that are similar to cellular automata [6] – particularly if the algorithms used at individual nodes are relatively simple. (Indeed, one can argue that given the limited, local knowledge available to a given node that complex algorithms, such those from modern control theory, will yield little by way of performance improvements. The other alternative, global knowledge, implies centralized control, which is inimical to the network architectures here.) The field of cellular automata studies how the overall system evolves given particular node behaviors within a discrete space-time system. One potentially useful aspect of the field of cellular automata, apart from self-organizing behavior, is that fairly sophisticated, coordinated global behavior can arise from these highly simplistic, locally interacting nodes [7], and the behaviors can be changed according to the network status simply by changing the CA (cellular automata) rules. A network coordinated by these global behaviors have the scalability and robustness that would otherwise be difficult to achieve in more centralized approaches. An example of the application of cellular automata to a related problem (the operation of location based services in a mobile network) is given in [8].

- *Swarm intelligence.* Biological swarms (e.g. ant colonies) are a good example of self-organized systems based on distributed processing. As such investigations into the mechanisms used to regulate their operations may provide useful templates for wireless networks. For example, consider the principle of stigmergy [9], a mechanism of coordinated behavior within a swarm whereby modifications of the environment by one member of the swarm results in changes in the behavior of other members. This has direct analogies with the adjustments needed among radio access nodes with limited direct intercommunication: often a change in the configuration of one access node may be only perceived by other access points by its impact on the radio environment and associated mobiles.
- *Microeconomics of oligopolies and game theory.* This is one of the best-known paradigms for the design of distributed systems competing for resources, due to the recognized optimizing properties of free market scenarios [3]. Examples of distributed algorithms using the concept of an *abstract* "market price" for a given resource include those given in [4] for call routing and in [5] for wireless ad-hoc networks. However, further work is needed to use such concepts in a more holistic approach, taking account of the *actual* economic drivers.

It should be noted that there are close interrelationships between the above frameworks, but they do provide distinctly different perspectives that should be explored. For example, there are equivalences between auctions in free-market economics and response threshold models in swarm theory [9] and yet a competitive market environment has different drivers from those in the collaborative structure of a social ant swarm. Both sets of drivers are to be seen in radio access networks – collaboration is needed within a network; competition, between the networks of different operators.

While the above theories can be used both for algorithm design and for the analyzing the resulting performance, there are yet other areas that can be used for analyzing the resulting behavior and performance:

- *Spatiotemporal models of population growth* [10]. The deployment of a wireless network in the presence of competition from other operators (using the same or different technologies) is directly analogous to the growth of a population of a particular species competing with others for limited resources. In this case, the resources are not spectrum but end-users. Therefore, mathematical models of ecologies should have direct relevance for the prediction of network growth. In particular, niche theory, with its concept of partially overlapping niches [11], can be seen as excellent framework for analyzing the impact of competition of systems using different technologies (e.g. third-generation cellular versus WLAN) specializing in different, but overlapping, ranges of service types.
- *Entropy-based complexity measures.* Such measures [12], [13] can be used to characterize behavior, and hence performance, in complex systems. This type of approach was recently used in [14] to control the configuration of transmit power levels in wireless networks and as such, represents a good example of how an analysis technique can be then used to drive the system design. Furthermore, the global behaviors of the complex systems considered here are prone to instability due to phase transitions [15], [16]. Hence, for a network operating with

distributed, interacting and autonomous nodes to be used with confidence, there is also a need for a mechanism to detect and avoid such critical points within the network: an entropy-based complexity measurement is one such mechanism.

Fundamentally, wireless systems are rapidly approaching the complexity of natural systems and have similar drivers of competition, collaboration, limited resources, etc. Hence, the design of the algorithms for their configuration and deployment should co-opt any insights available from the bodies of science already available from the worlds of biology, automata, and economics. However, for the particular application space here, namely the design of algorithms for a self-deploying, self-configuring wireless access network, advances are needed in the basic sciences. In particular these are far from maturity on the synthesis side. Analytical techniques yield important information regarding existing systems, but it is synthesis that is a key requirement for the ability to engineer solutions.

4 Cellular Automata: A Case Study

Recent work [14, 17] shows an example of the use of a two-stage, CA-like algorithm for the auto-configuration of base stations' pilot transmit power levels. By adjusting these, the method aims to achieve the best coverage in an area, but using a distributed algorithm at each base station and only localized information of neighboring base stations within its range. The base station is given several states, where at each state, the base station will perform certain functions. Which state the base station is in is determined by a set of CA-like transition rules, which changes the state of the base station depending on the states of its neighbors. In Figure 3, there are 100 base stations, placed roughly 2500 meters apart in a region with uniform radio propagation conditions, and deployed at random times. The base stations would, upon deployment, enter a state when seeks out its surrounding neighbors and approximate their distances, and sets its cell size accordingly by adjusting its power levels.

Once that state is done, the base stations enter a second stage when it uses feedback from the mobiles to make minor adjustments to the cell sizes to fill up the smaller gaps in the coverage, before eventually settling into a stable, static state. During this second stage, the base station keeps track of the mobiles that are connected to it. Each mobile monitors the signal it receives from the base station that it is connected to. When the signal begins to go below a predetermined threshold and the mobile cannot find a signal from a neighboring base station, the mobile sends out a signal to the base station to which it is connected to indicate a possible gap in coverage. When a mobile reports a possible gap in coverage, the base station to which it is connected increases its cell size by an increment. Periodically, the base station checks the status of the mobiles in its cell and increases its cell size by an amount that depends on how many mobiles have reported gaps in coverage. The cell size is increased by a factor of F:

$$F = ne^{-2d}$$

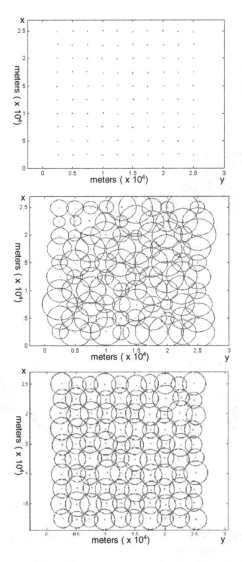

Fig. 3. Graphical representation of the coverage areas of a set of base stations (from top to bottom) before deployment, during deployment, and after deployment

where n is the number of mobiles that have reported coverage gaps and d is the difference between the current cell size and the cell size that was established during the initial deployment stage. This factor ensures that a given base station does not increase its cell size too much and that both the base station and its neighbors increase their cell sizes evenly to cover the gap. Figure 4 shows the number of mobiles that were dropped due to gaps in coverage from the start of the second stage, which is reduced over time as the gaps are eventually covered.

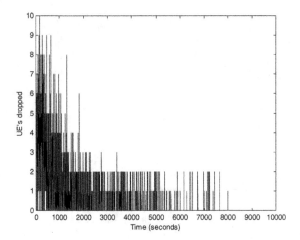

Fig. 4. Number of UEs dropped over time because of gaps in coverage

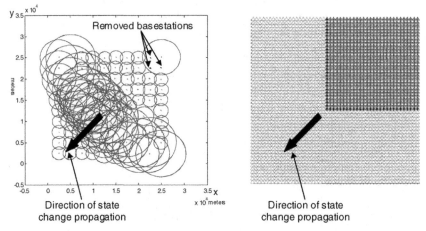

Fig. 5. Propagation of state changes in cell-boundary view (left) and 2-d CA view (right) when a number of base stations are removed

This CA-like arrangement also enables reconfiguration whenever a base station is added or, as shown in Figure 5, removed. When changes are made to one part of the network, this would trigger state changes in neighboring base stations, and this change then propagates throughout the entire system, prompting every base station to readjust to compensate for the addition or removal of base stations. Figure 5 also shows the implementation of the algorithm's state transition rules in a more recognizable 2-dimensional CA, showing the state change propagation in CA form. The mapping of the system to the 2-dimensional CA is achieved by treating each base station as a CA cell, each having the same three states and transition rules as the base stations. The neighborhood list of each cell in the CA reflects the interactions between the base stations, and the size of the list is dependent on the maximum range of the base station. This illustrates the advantages of achieving flexible, decentralized control by using simple, localized CA rules, demonstrating the potential of this field within the application domain of wireless access networks

5 Conclusions

This paper has examined the current and future trends in next-generation wireless access networks that will lead to the increasing significance of the costs associated with the deployment and configuration of such networks. To address this, the concept of a self-deploying, self-configuring radio access network was proposed. The sciences of complex systems, whether they come from economic theory, ecology/population growth models, or cellular automata should be capable of providing solutions for the required algorithms. An example, taken from the field of cellular automata for a radio network capable of self-adaptation to achieve universal coverage in a simplified environment, was examined.

While there are significant challenges posed by the vision here, if the appropriate robust, synthesis techniques can be found then self-aware, self-designing radio access networks will result. These will enable additional complexities at the radio-system level to be accommodated without overwhelming the system owner with an infeasible uphill struggle to find an efficient deployment. This could well be the difference between widespread adoption and economic non-viability of next-generation wireless access architectures.

Acknowledgements

A portion of this work was part-supported by the EU Commission through the IST FP5 project IST-2001-39117 ADAMANT.

References

[1] Wireless World Research Forum, "The Book of Visions 2001: Visions of the Wireless World", http://www.ww-rf.org/general_info/bookofvisions/bov.html, 2001.
[2] WWI Ambient Networks Project, IST Project Number 507134, http://www.ambient-networks.org/.
[3] C. Courcoubetis and R. Weber, *Pricing Communication Networks*, Wiley, Chichester, UK, 2003.
[4] M. Gibney, N. Jennings, N. Vriend, and J. Griffiths, "Market-Based Call Routing in Telecommunications Networks Using Adaptive Pricing and Real Bidding" in *Agent Technology for Communication Infrastructures*, ed. A. Hayzelden and R. Bourne, Wiley, Chicester, UK, 2001.

[5] V. Rodoplu and T. Meng, "Core Capacity of Wireless Ad Hoc Networks" in *5th International Symposium on Wireless Personal Multimedia Communications*, 2002, vol. 1, pp. 247–251, 2002.

[6] P. Sarkar, "A Brief History of Cellular Automata," *ACM Computing Surveys*, vol. 32, no. 1, pp. 88–107, March 2000.

[7] S. Wolfram, "Universality and complexity in cellular automata", *Physica D*, vol. 10, no.1-2, pp. 1-35, Jan 1984.

[8] R. Subrata and A. Zomaya, "Evolving Cellular Automata for Location Management in Mobile Computing Networks," *IEEE Transactions on Parallel and Distributed Systems*, vol. 14, no. 1, Jan. 2003.

[9] E. Bonabeau, M. Dorigo, G. Theraulaz, *Swarm Intelligence: From Natural to Artificial Systems*, Oxford University Press, 1999.

[10] F. Hoppensteadt, *Mathematical Methods of Population Biology*, Cambridge University Press, 1982.

[11] P. Nijkamp and A. Reggiani, "Non-linear evolution of dynamic spatial systems: The relevance of chaos and ecologically based models," *Regional Science and Urban Economics*, vol. 25, pp. 183–210, 1995.

[12] R. Lopez-Ruiz, H.L. Mancini, X. Calbert, "A statistical measure of complexity," *Physical Letters A*, vol. 209, pp. 321–326, 1995.

[13] J.S. Shiner, M. Davidson, P.T. Landsberg, "Simple measure for complexity," *Physical Review E*, vol. 59, no. 2, pp. 1459–1464, 1999.

[14] Ho L., Self-Organising Algorithms for Fourth Generation Wireless Networks, Ph.D. Thesis, Queen Mary and Westfield College University of London, Nov. 2002.

[15] P. Bak, C. Tang, K. Weisenfeld, "Self-organized criticality", *Physical Review A*, vol. 38, no.1, pp. 364-372, July 1988.

[16] Krishnamachari B., Bejar R., Wicker S., "Phase transitional phenomena in wireless ad-hoc networks," *Symposium on Ad-Hoc Wireless Networks, Globecom 2001*, November 2001.

[17] L.T.W. Ho, L.G. Samuel, J.M. Pitts, "Applying Emergent Self-Organizing Behaviour for the Coordination of 4G Networks Using Complexity Metrics", *Bell Labs Technical Journal*, vol.8, no.1, pp.5-26, 2003.

Content Distribution Through Autonomic Content and Storage Management[*]

Nikolaos Laoutaris, Antonios Panagakis, and Ioannis Stavrakakis

Department of Informatics and Telecommunications,
University of Athens, 15784 Athens, Greece
{laoutaris, apan, ioannis}@di.uoa.gr

Abstract. Content Distribution has to date been addressed by a mix of centralized and uncoordinated distributed processes, such as server replication and traditional node caching mechanisms, respectively. It is an inherently distributed process that is also increasingly relying on entities that are not only increasingly distributed but also increasingly autonomous. Consequently, centralized – and typically targeting the "socially optimal" – decisions are rather unrealistic for a distributed environment of autonomic entities. Instead, a distributed management of the engaged autonomic entities, which take decisions dynamically, should be key to efficient content distribution. The latter is advocated in this paper in which two entities that are central to content distribution - specifically the content and the node storage – are considered and it is discussed how their autonomic behavior drives the operation of a content distribution network. In the first case, it is the content that manages itself by dynamically generating duplicate copies and pushing them to (seizing) the appropriate storage. In the second one, it is the node storage that is in charge, deciding on the content to be locally stored. The decisions taken by the distributed and autonomic entities may – in the extreme case – be driven by self-awareness and self-interest only, without any network state information and co-operativeness. Or, they may use (some) network information and take decisions in a more cooperative manner, despite their autonomic and self-interest-driven nature. An example is presented on the later case, showing the potential both social and individual benefits.

1 Introduction

Communication networks have up to very recently been designed, optimized, and built, based on a careful planning and allocation of the primary network resource, the bandwidth. The emergence of the Internet and the World Wide Web as the main information delivery vehicles of our society, have necessitated the deployment of large amounts of network storage (or memory). The addition of storage capacity in network nodes, for the caching or replication of popular information documents in

---

[*] This work and its dissemination efforts have been supported in part by the IST Program of the European Union under contract IST-6475 (ACCA) and by the project entitled "Always Best Connected Provision in Heterogeneous Mobile Networks" funded by the Greek Ministy of Education under the framework "Pythagoras".

M. Smirnov (Ed.): WAC 2004, LNCS 3457, pp. 69–78, 2005.

close proximity to the end users, has appeared as a viable and efficient alternative to adding more bandwidth, or deploying complicated quality of service architectures. It is generally believed that the deployment of network storage has helped in reducing end-user delays, network traffic, and in improving the overall scalability in the Internet content distribution chain. Thus, network storage has emerged as another important network resource that can substantially enhance the network performance and/or reduce the requirement for bandwidth.

Following the usual Internet development track, storage capacity has been added progressively by a plethora of different authorities and applications that, in most cases, operate independently. Services and applications like CDNs, P2P, and others like end-system multicast, have formed logical overlay networks over the physical internet infrastructure. As a result, the Internet has been seeded by a large amount of storage capacity that now serves as a common substrate for the support of a diverse set of content delivery schemes, both contemporary and planned ones.

Despite the impressive reduction in the cost of storage brought by the latest generation of storage devices, storage remains a valuable resource (both in owning and also in managing), especially in view of the latest user trend to exchange voluminous information documents, e.g., multi-megabyte music and video files which, by latest reports, amount to well above of 75 % of all Internet traffic [19]. Contributing to this trend is also the automatic dissemination of software updates (operating system/application updates, virus fixes) that has become a standard feature of most operating systems and applications [16]. The combination of uncoordinated deployment of storage, and the conventional wisdom that storage is cheap, has resulted in a rather limited emphasis on exploiting the new resource up to its full potential, and has set the stage for what appears to be a new contention for resources - this time for storage capacity - fuelled by the desire to disseminate voluminous content.

If the provisioning of memory continues to materialize as it has in the recent past, then in the very near future (autonomous) memory pools (CDN nodes, or local proxy servers) will be in place in most systems that constitute the Internet [6]. Building adaptive overlay content distribution systems on top of the underlying memory pools can provide for a significant alternative to the static provisioning of memory as materialized with the current replication schemes that employ very large granules of memory (e.g., entire mirror site).

Should (autonomic) memory pools exist and be marketed, content creators (or intermediaries) can build distribution systems on them by leasing storage capacity dynamically. The main advantage of such a scenario is that memory will be utilized more efficiently, and at a finer granularity, as potential users or applications will be able to use it on-demand and release[1] it when no longer needed making it available to other users that may request and pay for it (protocols and e-currencies for such resource trade paradigms have been proposed recently

---

[1] Datagrams (on demand allocation of bandwidth to packets) has been the cornerstone of the Internet. The on demand allocation of storage to content seems to be as meaningful.

[21]). Re-organizing the memory is not possible with the current installment of dedicated mirrors and proxies in fixed locations and with fixed capacities. It is believed that the ability to reorganize the existing resources will be central to future intelligent information systems (see IBM's *autonomic computing* initiative [8]). The decisions concerning the management of the resources should be based both on the content requirements and the storage availability. In the sequel we take two different approaches at discussing the efficient utilization of the storage resource, one from the perspective of content itself, and the other from the perspective of the amorphous storage.

Our motivation for discussing these two approaches is to augment the current paradigm of placing the content only at fixed distribution points (e.g., the point of emergence (creation) and some mirror points), by allowing for the content to track adaptively the topology of the demand, and initiating a migration towards the areas of high demand without the intervention of a centralized authority. Such an approach will hopefully allow for a group of cooperating nodes to adaptively track and best serve the demand, without requiring centralized control; such a control is usually not present in Internet applications that are distributed and handled by multiple authorities.

2   Content Perspective

In this section we examine some requirements for autonomic content distribution, stemming from the perspective of content; the amount and the locations of the available storage are considered to be known here. This could materialize by first communicating with a Storage Broker entity (centralized or distributed one) from which storage is leased dynamically. The goal is to use the available storage to best serve the dynamically changing demand. For this purpose we conceptualize the tools of content movement and content duplication.

Content movement aims at pushing the appropriate content closer to the appropriate location. Content duplication spawns dynamically multiple copies of an information document in accordance with the request intensity; high demand leads to the increase of the number of copies, while a low demand leads to the decrease of the number of copies, trying to maintain an appropriate number of copies at various locations, to best suit the demand from the clients. We can make the following observations regarding the essence of employing each one of these concepts in isolation.

Sole application of content duplication without a limit on the number of replicas (extreme self-serving content behavior) for a given document allows multiple (or even all) clients to "own a local copy" of the desired document. A common problem with this strategy is that – due to the lack of coordination and the unrestricted number of allowed replicas – it leads to an excessive repetitious replication of the same (few) documents; the latter is clearly sub-optimal, considering that an off-line optimal replication policy forces multiple clients to "share" a single document replica, thereby increasing the number of distinct documents that may be hosted altogether.

Sole application of content movement with a small-fixed number of document replicas available, leads to a game of "tug-of-war", where the client (individual or group) that issues the most requests for a document succeeds in drawing it closer. Although it is justifiable to have the content closer to the location of highest demand, content movement alone falls short of best handling the demand, as it has to operate under a fixed number of document replicas, which may be a serious restriction. Under a fixed number of replicas that happens to be lower than the number of high-demand locations, the users will be served under a sub-optimal solution, as the freedom of allowing each location to have its own local copy (if that were the optimal solution under high enough request rates) would not exist.

The above discussion suggests that both concepts be employed and an adaptive trade-off between content duplication and movement be exercised by an efficient content distribution strategy. Thus practical and efficient rules for regulating between "tug-of-war" (forcing the clients to share) and "own local copy" (allowing multiple clients to have local copies when the corresponding request rates are high enough, which in turn limits the number of hosted distinct documents) should be identified. This fundamental trade-off between the number of replicas of each hosted document and the number of distinct documents is at the heart of an efficient utilization of the storage resource.

An interesting possibility is to consider autonomic content entities by assigning the responsibility for movement and duplication decisions to the content itself, rather than the content creator and the origin server that first injects the content to the network. To be able to make such decisions, content must be accompanied by a set of attributes that will allow it to act in an autonomic manner. As an example, imagine a movie file that is being injected in the network from the location of its origin server. The creator of the movie supplies it with attributes like storage credits (i.e., a budget for buying storage at replication points), maximum lifetime, "geographical" boundaries (set of ISPs in which it may spread) and other general characteristics for empowering its ability to manage itself. One such ability is the ability to split itself. This is stimulated by an interesting categorization of the targeted content among integral documents (i.e., documents that are indivisible, one document=one file) and non-integral documents (one document divided into multiple parts). The first case to be used with small/medium sized documents (e.g., html pages, images) while the second to be used with voluminous documents (e.g., software updates). The case of voluminous non-integral documents calls for special handling as compared to the case of integral documents (whose relatively small size permits them to move and duplicate as a whole). For voluminous content, applying movement and duplication to the entire document might be restricting due to large size; this is because potential movement and duplication actions become infeasible as few hosting nodes can accommodate such large object in their entirety. Segmenting such documents into multiple parts, that may be handled by different nodes, partly alleviates the problem of volume, but creates new challenges for orchestrating among the different constituent parts. In such cases, additional rules must be defined so as to maintain some degree of coordination among the multiple parts, as they move and duplicate about the network in response to the demand. A foreseeable target for such a coordination is, for example, to guarantee that the multiple parts constituting an entire document remain within a maximum distance of each other, so as to facilitate an uninterrupted (parallel or sequential) streaming towards a receiver.

Off-line algorithms that have a priori knowledge of the demand and topology are able to optimize the trade-off between the aforementioned concepts of *content movement* and *content duplication* by computing the relative value of each additional replica of a document and balancing it against the relative value of hosting a new (not yet replicated) document. Achieving this optimal trade-off in a distributed, on-line manner is challenging and yet unexplored and could be pursued by the proposed combination of the proposed concepts of content movement and duplication, possibly including additional recently proposed ideas such as parallel downloads and appropriate encoding schemes.

3   Storage Perspective

In this section we discuss the role of storage in an autonomic content distribution framework. We assume that storage is employed by the nodes for replicating content so that they may provide it to local users promptly, while limiting the consumption of bandwidth; essentially, the installed storage is used for absorbing the local demand for content, and not letting it flow to the network. Traditionally, a node's storage may be either managed by a central authority (e.g., owner of a CDN) in a way that maximizes the network's benefit, or by the individual node in isolation (e.g., typical user caches), in a way that maximizes the specific node's benefit.

The huge proliferation of the installed interconnected node storage calls for a reconsideration of the aforementioned traditional storage management paradigms. On one hand, centralized decisions are less feasible due to the lack of a single owner of the resources. On the other, the autonomous node storage facilities should not be managed in isolation catering to their own needs in a selfish and myopic manner but, instead, cooperation among the otherwise autonomic node storage entities should be considered.

The way that nodes cooperate in utilizing their storage resource is ultimately shaped by the scope of their utility, whether local (selfish behavior) or global (social aware behavior). We discuss such issues using the abstraction of a distributed replication group [14]. Such an abstraction is commonly employed for studying content distribution application such as web caching, web mirroring, content distribution networks (CDNs), and peer-to-peer applications.

Under this abstraction, nodes utilize their storage capacity to replicate information objects that they make available to local and remote users. A request that is issued by a local user and can be serviced locally (i.e., it involves a locally replicated object) is served immediately thus incurring a minimal cost. Otherwise, the requested object is searched in other nodes of the group and if not found, it is retrieved from the origin server; the access cost, however, increases with the distance. Depending on the particular application, the search for objects at remote nodes may be conducted through query protocols [22], succinct summaries [5], DNS redirection [17] or distributed hash tables [20].

Several placements problems can be defined regarding a distributed replication group. The proxy (or cache, or mirror, or surrogate) placement problem has been studied in several works, including [15, 18, 4]. The object placement problem refers to the selection of objects for the nodes, under given node locations and capacities [14, 10, 9, 2]. Finally, works such as [12, 13] combine node placement, node dimensioning and object placement in one problem.

All the aforementioned work in the field has centered around the optimization of the so called *social utility*, which is made of the sum of the individual *local utilities* of the nodes; here the term utility refers to delay and bandwidth gains from employing replication. The quest for optimizing the social utility arises naturally in applications where a centralized authority dictates replication decisions to the nodes. It suits well applications such as web mirroring and CDNs, which are operated under centralized control (the content creator or content distributor playing that role). Applications that are run by multiple authorities, such as web caching networks and P2P networks may too adhere to the goal of an optimized social utility, but this may come only as an act of voluntarism, as the optimization of social utility is often harmful to several local utilities.

Take as an example a group of nodes that collectively replicate content. If one of the nodes is generating the majority of requests, then a socially optimal (SO) object placement ends up using the storage capacity of other nodes to replicate objects that do not fit in the over-active node's cache. The users of these other nodes experience a service deterioration as a result of their storage being hijacked by potentially irrelevant content; in fact, such nodes are better off acting on their own, and employing a greedy local (GL) placement (i.e., replicating the most popular objects as pertaining to the local demand). The same situation arises if caching, rather that replication, is in place; remote hits originating from other nodes may evict objects of local interest in an LRU operated cache that participates in a web caching network. Fear of such exploitation may prevent nodes from participating in such groups and instead lead them to operate in isolation in a greedy local manner.

Being greedy local is often ineffective not only to the social utility, but to one's local utility too. For example, when nodes have similar demand patterns and inter-node distances are small, then replicating multiple times the same most popular objects, as done by a GL object placement, is highly ineffective. Clearly, there is a substantial gain for all nodes in that case, if they cooperate and replicate different objects; in fact, all local utilities may improve as compared to the GL performance, if such cooperation takes place. Nodes, however, are generally not aware of the remote demand patterns, thus cannot recognize such opportunities for cooperation. On the other hand, as discussed earlier, they cannot blindly trust a SO object placement as they do not know whether it will be for good or bad as pertaining to their local utility.

To address such problems equilibrium (EQ) placement strategies, which can guarantee that a node's local utility will always be better under EQ than under GL, may be used. A node has no reason not to participate in such placement strategies, as it has only to benefit from such participation.

For example in [11] an EQ strategy has been presented, which is based on the notion of Nash equilibrium, and extends the replication problem defined by Leff et al. [14] (who have developed the SO replication strategy) to the case of multiple local utilities. A two-step local search (TSLS) algorithm is derived that computes the EQ strategy. The TSLS algorithm can be implemented in a distributed manner, and for its execution each node needs to know only its local demand pattern and the objects selected for replication by remote nodes, but not the remote demand patterns (as required by centralized replication algorithms that compute the SO strategy). In addition, a distributed protocol that implements TSLS and requires minimal exchange of information has been developed. In the sequel we give a numerical example with the aim of highlighting the aforementioned placement strategies and their relevance to the self-interest of individual nodes.

3.1   A Numerical Example

In this section we give a numerical example to demonstrate the potential benefits of the TSLS algorithm. This is an example of an algorithm that is run by each autonomic node in order to take decisions in a cooperative (not isolated) framework utilizing (some) limited network information and yielding decisions that increase the global gain (benefit) without ever reducing the individual gain (benefit) enjoyed when acting in a self-serving manner only, in isolation.

There are two nodes that generate requests from the same Zipf-like distribution that assigns to the ith most popular object a request probability K/i^a, where $K = (\sum_{i'=1}^{N} \frac{1}{i'^a})^{-1}$; N denotes the number of distinct objects, a the skewness parameter of the distribution, and ρ_j the total request rate from the jth node (here $j = 1$ or 2). The local access cost is, $t_l = 0$, the remote one, $t_r = 1$, and the cost of accessing the origin server, $t_s = 2$; this leads to a hop-count notion of distance. It is assumed that there exist $N = 100$ distinct objects, and that each node has a storage capacity for $C = 40$ objects.

In Table 1 we show the objects replicated under the GL, SO, and EQ replications strategies for fixed $\rho_1 = 1$ and varying ρ_2. The GL strategy selects for each node the first 40 most popular objects, i.e., those with ids in $\{1:40\}$, independently of ρ_2. The SO strategy, however, is much different. As the request rate from Node 2 increases, SO uses some of the storage capacity of Node 1 for replicating objects that do not fit in Node 2's cache, thereby depriving Node 1 of valuable storage capacity for its own objects. For $\rho_2 = 10$, Node 1 gets to store only 3 of its most popular objects, while it uses the rest of its storage for picking up the next 37 more popular objects for Node 2, starting with the one with id 41. Under the EQ strategy Node 1 (v_1) stores 23 of its most popular objects. Node 2 (v_2) is the second one (i.e., the last one) to improve its placement, and it naturally selects the initial 40 most popular objects.

Table 1. An example with v_1, v_2 having the same Zipf-like demand pattern with $a = 0.8$. The number of available objects is $N = 100$ and the storage capacity of each node is $C = 40$. Also, $t_l = 0$, $t_r = 1$, $t_s = 2$, $\rho_1 = 1$

placement strategy	Node 1 objects	Node 2 objects
GL, $\rho_2 = X$	$\{1{:}40\}$	$\{1{:}40\}$
SO, $\rho_2 = 1$	$\{1 : 16\} \cup \{41 : 64\}$	$\{1{:}40\}$
SO, $\rho_2 = 2$	$\{1 : 12\} \cup \{41 : 68\}$	$\{1{:}40\}$
SO, $\rho_2 = 3$	$\{1 : 9\} \cup \{41 : 71\}$	$\{1{:}40\}$
SO, $\rho_2 = 4$	$\{1 : 7\} \cup \{41 : 73\}$	$\{1{:}40\}$
SO, $\rho_2 = 5$	$\{1 : 6\} \cup \{41 : 74\}$	$\{1{:}40\}$
SO, $\rho_2 = 6$	$\{1 : 5\} \cup \{41 : 75\}$	$\{1{:}40\}$
SO, $\rho_2 = 7$	$\{1 : 4\} \cup \{41 : 76\}$	$\{1{:}40\}$
SO, $\rho_2 = 8$	$\{1 : 4\} \cup \{41 : 76\}$	$\{1{:}40\}$
SO, $\rho_2 = 9$	$\{1 : 3\} \cup \{41 : 77\}$	$\{1{:}40\}$
SO, $\rho_2 = 10$	$\{1 : 3\} \cup \{41 : 77\}$	$\{1{:}40\}$
EQ, $\rho_2 = X$	$\{1 : 23\} \cup \{41 : 57\}$	$\{1{:}40\}$

We turn our attention now to the average individual and social access costs under the various placement strategies. The local access cost for node v_j is $\sum_{o_i \in P_j} r_{ij} \cdot t_l + \sum_{\substack{o_i \notin P_j \\ o_i \in P_{-j}}} r_{ij} \cdot t_r + \sum_{o_i \notin (P_j \cup P_{-j})} r_{ij} \cdot t_s$, whereas the social cost is the weighted sum of access costs of individual nodes, the weighing factor being the normalized request rate $\rho_j / \sum_{v_{j'} \in V} \rho_{j'}$; r_{ij} denotes the request rate at node v_j for object o_i, P_j denotes the *placement* of node v_j, i.e., the set of objects that it replicates, whereas P_{-j} denotes the collective set of objects that are replicated at all nodes except v_j. Figure 1 shows that as ρ_2 increases, the access cost for v_2 under SO decreases as it intercepts storage from v_1 for replicating objects according to its preference; v_1's access cost under SO increases rapidly as a result of not being able to replicate locally some of its most popular objects. In fact, for $\rho_2 > 2$, v_1's cost is worse (higher) that the corresponding one under GL. From this point and onwards, v_1 is mistreated by the SO strategy and thus it has no incentive in participating in it, as it can obviously do better on its own under a GL placement. Notice also that as a consequence of v_2's higher request rate, the social cost under SO follows in profile v_2's cost under SO.

By following the EQ strategy, v_1's cost cannot become higher than under GL, that is, v_1 cannot be mistreated, independently of ρ_2 and other parameters. In fact, both nodes succeed in doing better under EQ than under GL. Node v_2, however, benefits the most, and thus incurs a lower cost than v_1. This owes to the fact that v_2 is the second (last) one to improve its placement and, thus, has an advantage.

4   Related Work

We are aware of only few very recent works on game-theoretic aspects of caching and replication. Hadjiefthymiades et al. [7] (May, 2004), have studied the contention between different users that compete for storage in a single cache, and have

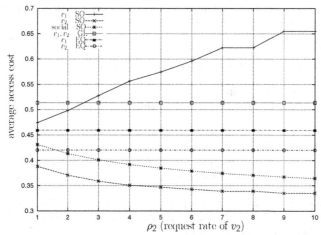

Fig. 1. Average cost for the example of Table 1: "v_j – XX" denotes the local cost for node v_j under the placement strategy XX; "social XX" denotes the social cost under the placement strategy XX

modeled it as a continuous game. More relevant to our work is the work of Chun et al. [3] (July, 2004), which studies distributed selfish replication. However, this work does not consider storage capacity limits on the nodes and, thus, differs substantially from our approach. Recent works on incentives in P2P networks, e.g., Antoniadis et al. [1], study the problem of attracting users to a P2P network and making them contribute more content. The aforementioned work, however, formulates the problem at a completely different level as compared to the current work, as it focuses on the number of files shared by each node, without identifying the identities of these files, whereas we focus on identifying the exact set of files shared.

References

1. Panayotis Antoniadis, Costas Courcoubetis, and Robin Mason. Comparing economic insentives in peer-to-peer networks. Computer Networks, 46(1):1–146, September 2004.
2. Ivan D. Baev and Rajmohan Rajaraman. Approximation algorithms for data placement in arbitrary networks. In Proceedings of the 12th Annual Symposium on Discrete Algorithms (ACM-SIAM SODA), pages 661–670, January 2001.
3. Byung-Gon Chun, Kamalika Chaudhuri, Hoeteck Wee, Marco Barreno, Christos H. Papadimitriou, and John Kubiatowicz. Selfish caching in distributed systems: A game-theoretic analysis. In Proc. ACM Symposium on Principles of Distributed Computing (ACM PODC), Newfoundland, Canada, July 2004.
4. Eric Cronin, Sugih Jamin, Cheng Jin, Anthony R. Kurc, Danny Raz, and Yuval Shavitt. Constraint mirror placement on the internet. IEEE Journal on Selected Areas in Communications, 20(7), September 2002.
5. Li Fan, Pei Cao, Jussara Almeida, and Andrei Z. Broder. Summary cache: a scalable wide-area web cache sharing protocol. IEEE/ACM Transactions on Networking, 8(3):281–293, 2000.

6. Garth A. Gibson and Rodney Van Meter. Network attached storage architecture. Communications of the ACM, 43(11):37–45, November 2000.

7. S. Hadjiefthymiades, Y. Georgiadis, and L. Merakos. A game theoretic approach to web caching. In Proceedings of IFIP Networking 2004, Athens, Greece, May 2004.

8. IBM Corp. Autonomic Computing initiative, 2002. http://www.research.ibm.com/autonomic/.

9. Jussi Kangasharju, James Roberts, and Keith W. Ross. Object replication strategies in content distribution networks. Computer Communications, 25(4):376–383, March 2002.

10. Madhukar R. Korupolu, C. Greg Plaxton, and Rajmohan Rajaraman. Placement algorithms for hierarchical cooperative caching. In Proceedings of the 10th Annual Symposium on Discrete Algorithms (ACM-SIAM SODA), pages 586–595, 1999.

11. Nikolaos Laoutaris, Vassilios Zissimopoulos, and Ioannis Stavrakakis. Distributed selfish replication, 2004. [submitted]

12. Nikolaos Laoutaris, Vassilios Zissimopoulos, and Ioannis Stavrakakis. Joint object placement and node dimensioning for internet content distribution. Information Processing Letters, 89(6):273–279, March 2004.

13. Nikolaos Laoutaris, Vassilios Zissimopoulos, and Ioannis Stavrakakis. On the optimization of storage capacity allocation for content distribution. Computer Networks, 2005. [to appear]

14. A. Leff, L. Wolf, and P. S. Yu. Replication algorithms in a remote caching architecture. IEEE Transactions on Parallel and Distributed Systems, 4(11):1185–1204, 1993.

15. Bo Li, Mordecai J. Golin, Giuseppe F. Italiano, Xin Deng, and Kazem Sohraby. On the optimal placement of web proxies in the internet. In Proceedings of the Conference on Computer Communications (IEEE Infocom), New York, March 1999.

16. J. Li, P.L. Reiher, and G.J. Popek. Resilient self-organizing overlay networks for security update deliver. IEEE Journal on Selected Areas in Communications, 22(1):189–202, January 2004.

17. Jianping Pan, Y. Thomas Hou, and Bo Li. An overview DNS-based server selection in content distribution networks. Computer Networks, 43(6), December 2003.

18. Lili Qiu, Venkata Padmanabhan, and Geoffrey Voelker. On the placement of web server replicas. In Proceedings of the Conference on Computer Communications (IEEE Infocom), Anchorage, Alaska, April 2001.

19. Stefan Saroiu, Krishna P. Gummadi, Richard J. Dunn, Steven D. Gribble, and Henry M. Levy. An analysis of internet content delivery systems. In Proceedings of the 5th Symposium on Operating Systems Design and Implementation (OSDI 2002), December 2002.

20. I. Stoica, R. Morris, D. Liben-Nowell, D.R. Karger, M.F. Kaashoek, F. Dabek, and H. Balakrishnan. Chord: A scalable peer-to-peer lookup protocol for internet applications. IEEE/ACM Transactions on Networking, 11(1):17–32, 2003.

21. David A. Turner and Keith W. Ross. A lightweight currency paradigm for the P2P reseource market, 2003. [submitted work].

22. Duane Wessels and K. Claffy. ICP and the Squid web cache. IEEE Journal on Selected Areas in Communications, 16(3), April 1998.

A Unified Framework for the Negotiation and Deployment of Network Services

Spyros Denazis and Lidia Yamamoto

Hitachi Europe SAS, Sophia Antipolis Laboratory,
1503 Route des Dolines, F-06560, Valbonne, France
{Spyros.Denazis, Lidia.Yamamoto}@hitachi-eu.com

Abstract. The Internet network technology today does not allow a sufficient degree of autonomy to express user choices, constraints and preferences in order to dynamically obtain the most suitable services. One of the goals of Autonomic Communication is to produce self-managing network elements able to provide the desired services in an automated way. In this context, we propose an architecture to automate user-provider and provider-provider relationships, by converting the Internet into an electronic market space where the commodities to be traded are network services. After an agreement has been reached via agent-based automated negotiation mechanisms, network elements must be automatically configured in order to enforce the agreed conditions. This is achieved by generating commands to programmable network elements via open interfaces. The ultimate goal is enable fully automatic installation, configuration and monitoring of protocols or service components involving multiple ownership domains, while taking into account the constraints and preferences of users and providers.

1 Introduction

The Autonomic Communication initiative [1] is investigating the inter-relation among network elements to understand how their behaviours can be learned, influenced or changed such that they can self-organize to provide the desired services in an automated way.

The Internet network technology today does not allow a sufficient degree of autonomy to express choices, constraints and preferences in order to dynamically select the most suitable services. The user typically has to undergo a series of manual steps, for example when trying to connect to a Wi-Fi hotspot, in which the user has to enter authentication and billing data. When roaming to a different domain, e.g. switching to a different hotspot, the user is most of the times obliged to enter the data again and decide whether to accept the service offer or not. When many offers are simultaneously available, the user must inspect each of them and make a choice. Roaming is thus most of the times not transparent, and when it is transparent (e.g. roaming between mobile telephone operators) it is not always guaranteed that the user will benefit from the best offer that matches his or her interests.

M. Smirnov (Ed.): WAC 2004, LNCS 3457, pp. 79–93, 2005.

We investigate the transition from the current Internet towards an Autonomic Communication Network (ACN). We focus on ways of bridging heterogeneity in requirements, interests, constraints and preferences.

We propose to model the new ACN as an electronic market space where network services are treated as goods that are traded among the different parties involved. We propose an open architecture able to make and enforce decisions even across ownership and administrative borders. This requires a unified framework for service description, announcement, discovery, negotiation and provisioning that takes into account the different preferences and constraints of the parties involved, and is able to achieve a common ground that is interesting for all parties involved. We use automated negotiation as the conflict resolution technique, and open interfaces for automated reconfiguration of services.

The same approach is applicable to several granularities of ownership borders: Users, groups of users, companies, network providers, etc. We will pay particular attention to user-provider (service provisioning) and provider-provider (inter-domain) interactions.

This paper is structured as follows: Section 2 provides a brief survey of the current state of the art in this topic, and identifies several open issues where more research is necessary. Section 3 presents a unified framework for creating an environment able to search for, negotiate, deploy, configure, monitor, reconfigure, and tear down end-to-end network services in an automated way across domain borders. Section 4 concludes the paper.

2 State of the Art

Today it is still difficult to get network services autonomously when cross-ownership interactions are required. This is the case for both the customer-provider case (e.g. fixed or wireless access) and the provider-provider case (inter-domain agreements). Nowadays these interactions still rely mostly on slow human communication (e-mail, fax, paper contracts, and so on). Solutions to automate specific parts of the process are available, and some of them are discussed in this section. However the integrated picture seems somehow still missing.

The challenge is to automate cross-ownership interactions in ACNs, in a way to accurately reflect users' preferences, providers' interests and concerns, as well as the multiple underlying network characteristics. Such interactions cannot assume cooperation from the communicating peers, therefore security and safety are major concerns. Considering the amount of already existing network providers and users, the potential pairs of interacting peers is large. In this context, interactions cannot rely exclusively on authentication and authorization as a security mechanism, since this would imply constraining the interactions to trusted peers. In this model, interactions are driven by the level of trust one peer places in another, and closely mimic the corresponding human interactions. When dealing with several levels of trust, binary access/deny mechanisms are not sufficient. Negotiation mechanisms are necessary to reach intermediate, compromise solutions (e.g. access to a certain amount of a resource, or to a specific part of a document).

Negotiation can be used for configuration set-up, information transfer, new service deployment, or the usage of some physical resource such as bandwidth, CPU, and memory. The parameters involved include quality requirements, performance level, prices, payment conditions, etc. The decision criteria relate to user preferences, for example, towards the fastest network access, or the closest, the most reliable, the cheapest, etc. The user could also prefer to combine service from multiple available providers in order to increase robustness.

2.1 Automated Negotiation

Automated negotiation [11,12] mimics human negotiation processes to reach agreements on one or more issues. It is an active research topic in the field of multi-agent systems, and has been applied to several areas including telecommunication and computer networks. A number of agent-based systems to enable provider selection and inter-domain interactions have been proposed [13,14,15,16,17]. The use of Agent Communication Languages enables rich and flexible interactions, which can be made interoperable through standardized specifications provided by the FIPA consortium [18]. Several FIPA standard protocols and languages are available that can be used for this purpose: Contract negotiation [19,20], brokering [21], proposals [22], auctions [23,24], QoS [25], network management [26]. Work is in progress towards a FIPA standard for agreement specification [27,28]. When ready it could be used, for instance, to specify a Service Level Agreement (SLA).

A multi-agent system for automated negotiation applied to VPN provisioning is described in [13]. Its agents comply with FIPA standards and implement multiple negotiation strategies. However it does not seek to optimise the VPN topology, resulting in a star configuration.

In the framework proposed in [16] the problem of inter-domain QoS routing is formulated as a Distributed Constraint Satisfaction Problem (DCSP). The QoS requirements considered are bandwidth and delay. A distributed algorithm is derived, which is suitable for unicast guaranteed QoS services. However, the algorithm is only valid when all domains on the end-to-end path support the specified agents and resource reservations, and cooperate to offer the requested service. In [17] the framework is extended with negotiation mechanisms and an ontology for VPN services. Price negotiation takes place after a path has been selected using the DCSP algorithm, therefore provider selection is not a direct outcome of such negotiation.

A formal model of the service selection problem is presented in [14], in the context of agent-mediated wireless access. A user agent called Personal Router acts on behalf of its owner to select wireless providers that better satisfy the user's preferences. These preferences are modelled as a utility function of receiving given service profiles. The selection problem is represented as a Markov Decision Process and the initial solution is to find those actions that maximize utility. However it is shown that the algorithm is computationally expensive. In [15] heuristic solutions are proposed that can make the problem more tractable.

2.2 Towards Automated Inter-domain Interactions

Automated inter-domain interactions are mostly limited to inter-domain route advertisement via Border Gateway Protocol (BGP) messages. Most ISPs today rely on the BGP community mechanism to have tighter control upon route propagation [4] by specifying preferred paths and deviating traffic outside their domains. There are many problems with this approach, as frequently pointed out in literature [5,6]: sub-optimal end-to-end paths, instabilities (route flapping), slow convergence in response to link failures. It would be better for the set of domains to cooperate in order to obtain the best routes according to given metrics that satisfy users' requirements. We are talking about inter-domain QoS routing. Although there has been some research in this topic [5] as well as IETF guidelines [7] it remains largely an open issue at the moment.

Proposals to include QoS information in BGP are presented in [8,9]. A signalling approach for network state management is proposed in [10] that can be used for intra-domain as well as inter-domain QoS and other monitoring and configuration tasks. None of these approaches is generic enough to express the complexity of fully competitive inter-domain interactions related to end-to-end services, in which the trade-off between competition and cooperation must be taken into account and quantified. A richer approach is needed to cover the whole service cycle in an end-to-end basis, including service request, negotiation, selection, set-up, monitoring, renegotiation, and tear down. Our ideas to achieve this goal will be discussed in Section 0.

There are other gaps in existing work towards automating interactions among multiple, potentially competing ownership domains. First of all, partial deployment must be supported. For instance, let us consider a path is made up of domains A, B, C and D in sequence, with A as source domain and D as destination domain. If a service uses providers A, B, C and D, but only providers A and C provide automated negotiation capability, then the network characteristics of providers C and D should be measured as a black box, such that some information is available in order to provide the customer with an estimation of expected service level. Although no absolute guarantees can be provided in this case, such estimation can represent valuable information to influence the customer's decision in favour of a given provider.

Cascade negotiation towards an end-to-end service is only partially supported in existing approaches. In the case of [16] the first domain agent (in the source domain) communicates with all the other domain agents on the path to a given destination. In the example of path A-B-C-D above, the agent in domain A would send negotiation messages to B, C, and D. It would be more transparent if A would negotiate with B, B with C, and C with D following the path sequence. Cascade negotiations are directly related to inter-domain routing: if a negotiation fails or if a provider fails to comply with agreements, an alternative provider could be selected, resulting in a different end-to-end path. In the same way as unilateral BGP policies may have negative impact on global routing, cascade negotiations could lead to routing instability if conducted in an ad hoc manner. Further research is needed to fully understand the impact of cascade negotiations on inter-domain routing, and to provide methods that can guarantee that a stable route is found in reasonable time.

2.3 Inter-domain Network Programmability Made Feasible

Programmable networks [36] have been proposed in response to the need of more flexible and customisable network nodes for improved services and faster service deployment. Using a programmable network infrastructure, applications can benefit from available processing time and memory storage in intermediate nodes, which could be used to install and execute customized service components. The use of such node resources within the network raises security and safety concerns, and can only be made realistic with tight security and resource usage control. In the wide area end-to-end case, the benefit of programmable networks might be realized only through the installation of customized components in several nodes potentially belonging to different administrative domains. This raises even deeper concerns as network providers will be more than reluctant to open their nodes to foreign code.

If we can design an automated negotiation mechanism which is rich enough to express the characteristics of dynamically deployable components to be installed in the network, including their provenance, resource consumption and reliability, this could encourage providers to allow trusted components to be installed and executed in the network nodes supplied for that effect, therefore stimulating the usage and deployment of new network services involving multiple domains.

2.4 Ubiquitous, Ad Hoc, Sensor, and Other Small-Device Networks

Parallel to what is happening to the Internet infrastructure, several infrastructureless, self-organizing networks are emerging, such as ad hoc, sensor networks, ubiquitous networking, home networks. These networks should be formed spontaneously anywhere at any time, without requiring network operators or network managers. ACNs could benefit from ideas stemming from these networks to help automating tasks such as network management and service provisioning. On the other hand, these light-weight networks will also need to connect to existing more complex network infrastructures, where most of the content and services can be found, such as the Internet, Intranets, and Virtual Private Networks (VPNs). Multiple alternative providers may be available at a given location (e.g. wireless and wired), and multiple terminals may be able to act as gateways from the ad hoc network to the outside infrastructure.

The users of a device network are then faced with the problem of which provider or set of providers to select for access to an infrastructure, and through which gateway nodes. The choice of an optimum or nearly optimum connection solution may be non-trivial, involving many parameters such as expected throughput and delay, price, eventual service guarantees, level of trust in known providers, etc. This should be handled in an automated way, such that the users simply specify their preferences and the network nodes cooperate to find the optimum solution.

Resources dedicated to device networks are often limited, such as low bandwidth wireless links, and small terminals such as PDAs and cell phones with slow CPUs small memory space, and short battery life. Such limitation might also open up new markets for infrastructure-based computational services targeted at complementing these resources by outsourcing or through a Grid-style distributed computing paradigm. The framework proposed in this paper can play a critical role in achieving this goal.

3   A Unified Framework for the Negotiation and Deployment of Network Services

In this section, we attempt to create a vision about the future Internet and some of the features it should manifest. According to it, we aspire to treat the Internet as a market place wherein the commodities are network services. The majority of electronic markets proposed so far, has been about tangible goods ranging from equipment to clothes, books, shares etc. In our case, we propose to treat the most basic service of Internet, namely, communication, as a commodity together with other related services and applications such as video conference, voice, etc.

Realising our vision and applying a methodology, we have borrowed from models describing interactions and operations taking place in a market economy leading us to a unified framework for service description, announcement, discovery, negotiation and provisioning. Such a framework also takes into account the different preferences and constraints of the parties involved.

3.1   Modelling Customer-Provider Interactions

The interactions that occur between end user and provider, and between peer providers can be modelled as producer-consumer interactions in a market economy. Numerous e-commerce systems have been proposed or are in use nowadays, which also model these economic agents.

Initial systems e-commerce systems [29,30] employed software agents as mediators for handling and automating interactions taking place in physical commerce. They followed Consumer Buying Behaviour (CBB) [31] to model actions and decisions that happen when buying and selling goods augmented to incorporate concepts from Software agents research. CBB models originating from traditional market research can be abstracted into one CBB model consisting of six stages that coarsely reflect consumer behaviour [31]:

- **Stage 1: Need Identification:** This stage is where customer realises his or her need for a specific product or service.
- **Stage 2: Product Brokering:** This stage answers the question of 'What to buy?'. The customer follows a course for gathering information in order to decide not only upon the product but also on its exact characteristics (product profile).
- **Stage 3: Merchant Brokering:** This stage answers the question of 'Who to buy from?'. The customer, having decided on the product profile, takes into consideration additional information about the merchant which are filtered through the customer's own criteria in order to reach a conclusion. For instance, lowest price, value for money, reputation, etc.
- **Stage 4: Negotiation:** This stage answers the question of 'How to buy?'. It is revolved around the rules governing the transactions between two parties. For instance, negotiating price or QoS level. This stage can be considered as a part of the previous two stages or a standalone stage depending on the type of market.

- **Stage 5: Purchase and Delivery:** This stage usually heralds the completion of the negotiation stage. It may also have an influence on the product and merchant brokering stages.
- **Stage 6: Service and Evaluation:** Finally, this is the stage where an evaluation period of the product and the promises that accompany it, commences. If such promises are not fulfilled the customer might decide to renegotiate or even choose alternative providers, going back to stages 4 or 2.

The CBB model and its stages above provide a very rough but reasonable guide for categorising the actions performed by any e-commerce system. Software agents materialise these actions in the context of the CBB model. Notably though, some of the stages often collapse into one, or overlap, and migration from one another can be non-sequential or iterative depending on the kind of e-commerce and eventually on the type of product(s) involved. Naturally, the variations of the CBB model have an immediate effect on the selection of agent technologies, languages, protocols, interfaces, and the actions agents perform which altogether constitute the e-commerce system.

Note however that the use of agent systems is restricted to the algorithms, concepts and protocols that are useful in a networking context, not the actual platform implementations. Agent platforms usually offer a complete infrastructure of services which is more suitable to support application-layer implementations. Network layer issues require lightweight methods that do not rely on an existing communication support - the framework itself is intended to provide such communication services, therefore cannot assume that they are ready for use. Ultimately, negotiation algorithms and protocols should be embedded as services into the autonomic communication system itself, at the same level of any other communication service also present there.

3.2 A High Level Description of the Unified Framework

The core idea behind our framework is the transformation of the Internet into an environment that acts as a distributed market place where potential merchants (Network or Internet Service Providers) and potential customers (home users, small enterprises, corporations, or even other ISPs) interact in order to compete and cooperate over selling, or buying a specific commodity, namely, *network services*. Starting with a competitive environment, the system should foster cooperation among providers in order to achieve improved end-to-end services.

Service Level Agreements (SLAs) play an important role in representing the profile of the product, i.e network services, in an unequivocal, and discrete manner. SLAs can, then, be used by the customers to quote a price, to negotiate with a provider(s), and evaluate the quality of service depicted in an SLA. In contrast, providers can compete with each other by lowering their prices over a requested SLA, or offering more advanced services on top of the requested ones, form pacts with other providers in order to increase their competitiveness, and reserve network resources for guaranteeing the SLA. Moreover, the collection of SLAs offers them a picture of their current and future resources needs, thereby facilitating management, re-engineering, and provisioning of infrastructure.

An SLA is an agreement between two roles, that of a customer and a provider. The customer may be an end-user or another provider. A retail SLA (r-SLA) is an agreement between an end-user and a service provider. A wholesale SLA (w-SLA) is an agreement between providers, and is usually based on aggregates meant to carry traffic from several end users. W-SLAs may be established in advance on a static basis, as part of a network provisioning phase, and independent of r-SLAs. However, a given r-SLA may also trigger the establishment of several w-SLAs across the domains involved, in a dynamic way.

Widespread use of SLAs aspires to establish a universal interface language among the involved parties representing services together with their characteristics that can be uniquely recognised and interpreted along the end-to-end path. Such language will facilitate the automation of network operations like customer-provider negotiations, (re)configuration of network resources etc. We expect that SLAs will form ontologies of objects ranging from generic ones customised for the needs of technology agnostic customers to detailed ones addressing the needs of experts like network managers.

Our framework should give the possibility to a customer to choose from a basic set of service parameters, like availability, throughput, latency, privacy, etc., those that are most desirable resulting in a *provisional SLA*. The user specifies high-level preferences and the user agent maps preferences to service parameters, then requests service from one or more providers via an automated negotiation mechanism. This activity (Fig. 1) can be considered as representative of *stage 2 (product brokering)* in the CBB model of the previous section.

The next step is to discover available providers and if necessary inform the customer about which providers may be contacted to participate in the negotiations. Providers could announce themselves with help of directories, beacons, or other mechanisms.

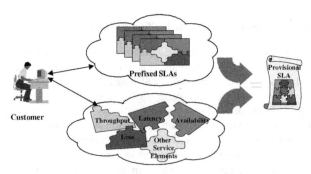

Fig. 1. Process of building Provisional SLAs

Using the customer service profile, the user agent must now decide upon the provider(s) to which to award the SLA. This is *stage 3 (merchant brokering)*. The decision may be completely automated or assisted by its human owner. Hence, the framework provides the means to support the submission of a provisional SLA to a number of providers along with an additional set of parameters that constitute the customer's criteria, e.g. price. This could be implemented using a standard language to express agreements (work is in progress in FIPA [27,28]). Receiving the provisional SLA and the customer's criteria, providers may respond with an offer on the submitted criteria or may even add more service elements or services in order to form an appealing package. This can be done through argumentation-based negotiation [11]. This offer returns to the customer in the form of a *Combined SLA* (Fig. 2). At this point the customer's agent (assisted or not by this customer) either selects a provider and accepts the provider's Combined SLA, or makes counter-offers. This is *stage 4* of the CBB model, the *negotiation stage*. Note that in this iterative negotiation model involving alternative providers, stages 3 and 4 are combined.

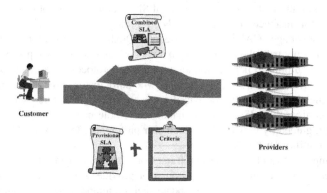

Fig. 2. Process of creating Combined SLAs

Taking the Combined SLA(s) for negotiation over some terms could be a complex and iterative process. For example, the customer's agent may request lowering the price in a gambling attempt, or by reducing some of the initial demands. Another possibility emerges when a final SLA has been agreed and the customer wishes to increase demand on some of the constituent parameters in the agreed SLA, i.e. more bandwidth. Providers may also make such offers when they see fit. It is expected that this stage will be involved in many temporal instances as the interactions between customers and providers occur dynamically. Each time that a new agreement takes place, the new terms are added to the existing SLA.

Given that an agreement has been reached between the two parties the delivery and charging mechanisms are triggered, entering thus *stage 5 (purchase and delivery)*. This corresponds to the configuration and operation of the service, together with an electronic charging and payment scheme. Furthermore, delivery requires from the provider to exert control on its infrastructure by reserving resources, applying scheduling policies, etc., in order to be able to deploy the services stated in the SLA. At the customer side this implies autoconfiguration of its network elements to support the desired service.

Finally, the last stage (*Service and Evaluation*) involves actions by both the provider and the customer. The provider should constantly monitor and manage its network so as SLAs from different customers are upheld, take correcting actions when needed, and policing customers that abide to the SLA. This requires inter-domain negotiations to reach agreements on which parameters a provider allows to be monitored by its customer or by peer providers. On the other hand, customers should also be given the ability to monitor the network so as to make their own choices. Out of this process, customers may either decide to modify some of the terms of the SLA in due course, returning thus back to the negotiation stage, or decide to form new SLAs for other set of services returning back to stage 2. Providers that fail to fulfil the requirements in an SLA risk sanctions or bad reputation. This requires infrastructure monitoring for connectivity and performance. In a general end-to-end case cascade monitoring might be needed as discussed in Section 2.2.

The idea of automating the negotiation of SLA parameters is not new, and has been presented, for instance, in the CADENUS project. However, to the best of the authors' knowledge, CADENUS focused on Quality of Service management and control. It did not treat the problem from a network autonomics perspective, nor did it make use of automated negotiation algorithms and protocols such as those discussed in Section 2.1. In CADENUS, it is the user who participates in the negotiation process by selecting the desired service based on a sorted list of offers, while automated negotiation algorithms (Section 2.1) handle the whole negotiation process on behalf of the user, guided by previously defined user preferences (expressed, for example, in the form of utility functions).

It is important to clarify that our framework does not dictate the way to express preferences and constraints at the customer or provider side, nor does it specify any feedback mechanisms to eventually update such preferences and constraints based on negotiation outcomes. For example, at the provider side, such preferences and constraints could be expressed using policy languages. In this case, we assume that quantitative parameters can be extracted from such policies, and that these parameters can be used to determine upper and lower thresholds in the negotiation process.

3.3 Automated Deployment and Configuration

Realising stages 5 & 6 requires technologies and network architectures capable of supporting automated deployment of the agreed network services. This also entails the (re)-configuration of the network elements across the different domains that the end-to-end path traverses. Furthermore, before or when the service has been deployed detailed monitoring data must be collected and processed (discussed in the next section)

Accordingly, there is a need for higher degrees of flexibility supported by networks and the network elements thereof. Such flexibility is measured as the network's capability to dynamically extend and change its overall functionality and behaviour through the on-demand introduction and configuration of functional components.

Engineering the (heterogeneous) network with this kind of flexibility requires the design and adoption of a number of key concepts proposed or extended by active and programmable networks initiatives [36]: service component models [33,34], open interfaces [38], and new network element architecture models [35,40].

Component-based models consider services as comprised of self contained building blocks that act as primitives that make up more complex services when combined in specific ways. Treating complex services as components increases the flexibility of their deployment and allows for better decision-making.

Open interfaces avoid dependencies on a small group of vendors created by proprietary interfaces. They allow algorithms and services to be developed independently of advances in the forwarding plane, as they allow seamless configuration of the control and management planes.

Network element and configuration models provide a common and unambiguous view of the network and its state thereby facilitating the communication among different parties. They also provide a richer set of information that can drive and enhance the decision-making and monitoring algorithms.

Collectively, these concepts constitute a universal network language in much the same way as SLAs represent what customers want and understand as network services. Through such language, providers are capable of representing and communicating with each other information about service entities, network resources and their location, implementation technologies, vendors' equipment etc. Based on this information they can then feed their decision making algorithms about where, when and how to deploy a service, manage it throughout its lifetime or reconfigure it according to customer changing requirements.

In this context, upon completion of the negotiation phase with the customer, a provider translates the SLA to its corresponding service component model in the form, say, of an XML schema, which is then processed by the service deployment algorithm [37]. The decision of where to deploy the components of the service depends on the network model that captures information about the available resources and implementation environments. This information is compared against service requirements, for instance, how much bandwidth is needed, the implementation environment that the service components require etc.

When a decision is made the service deployment mechanism is contacted in order to enforce the decision. The enforcement is facilitated by open interfaces that abstract a common set of deployment mechanisms implemented according to the implementation platform of the network element. The decision and deployment process enforced in the provider's domain may trigger a series of similar decisions and deployment operations made by other providers that eventually will form the end-to-end path across the different providers' domains.

We envisage that the deployment of a service will be comprised of two parts: the deployment of the QoS model that satisfies the resource requirements and the deployment of functional components that will process packets beyond store-and-forward processing and belong to the service. The QoS deployment heavily depends on open programmable interfaces too, like IEEE P1520 [38]or ForCES [40] that facilitate a suitable mapping between resource demands and configuration operations on the network elements independent of the vendor or platform of the network element.

The deployment of a service is followed by the management of the service in such a way that the obligations of both parties described in the SLA are fulfilled. To this end an intelligent monitoring system must collect and disseminate statistics and data to the interested parties. A special operation of the monitoring system is to feed the network management algorithms with alarms so that proactive actions must be initiated and carried out in an effort to stick to the agreed SLA as close as possible. Again, the adoption of common models and of open interfaces enables automating and expediting these tasks through an ambient interoperability layer.

3.4 Automated Service Monitoring

After service deployment and configuration, the service should be monitored to comply with SLA, to identify non-cooperative or misconfigured domains, etc. The ability to perform inter-domain QoS measurements is crucial to provide reliable and high quality services. Monitoring is required for troubleshooting, and automated set-up of monitoring tasks is the first step towards automated diagnosis and repair. However today, monitoring an arbitrary end-to-end path today is difficult and restricted, and the obtained information is very limited and inaccurate. A great obstacle against global-scale performance monitoring today is that network providers are not willing to share information about their networks, due to fear of eavesdropping by competitors, fear of attacks, and various business reasons.

It would be beneficial to have an automated way to dynamically express which parameters may or may not be monitored across domains, depending on trust levels among providers. An automated negotiation mechanism would enable the automatic set-up of measurement tasks across domain, while at the same time respecting providers' policies and restrictions. This would also act as incentive for cooperation, as providers that cooperate to offer monitoring results would be in a better position to offer higher quality services appreciated by customers. We have taken a first step in this direction [2] by proposing an automated negotiation framework for the dynamic set-up of network monitoring tasks across domain borders.

In [2] we proposed to apply automated negotiation techniques as a way to dynamically agree on which QoS parameters may be monitored across domains, depending on the resources available within each domain, the current network conditions, the trust levels among providers, and their respective policies and constraints, including security and privacy constraints. We have identified the potential protocols and strategies that could be applied, and mapped monitoring parameters to them. As a format for the exchange of requested measurement parameters, we have selected the Specification of Monitoring Service (SMS) [3] proposed for inter-domain monitoring. The SMS format is a document format which contains the necessary parameters to request inter-domain QoS monitoring tasks.

The resulting monitoring data must be used as feedback information for the autonomic communication control system, such that deviations from the expected service can be promptly detected and a system reconfiguration can be triggered when necessary. This remains largely an open issue in current networks, where the focus is on database storage and visual analysis of measurement results.

An example of how monitoring information could be used as feedback for decisions processes in Autonomic Communication would be to automate diagnose and repair of network problems (troubleshooting). The challenge is to perform this across multiple ownership boundaries, in order to achieve consistent end-to-end service. Thaler et al. [41] propose a generic architecture for distributed troubleshooting which also works in the inter-domain scenario. The architecture includes a protocol called Globally Distributed Troubleshooting (GDT), for automated problem and status reporting across different domains. Nevertheless, automating network troubleshooting remains a non-trivial problem.

4 Conclusions

This paper aspires to increase awareness among researchers for greater degrees of automation in the network, and to identify specific aspects that must be engineered into the network in order to achieve this. Automated negotiation algorithms and protocols can be applied between customers-providers and providers-providers, as the current static model is very restrictive and outdated to cope with the requirements of a truly autonomic network, which must detect and resolve conflicts of interest in an automated way.

With this in mind, an initial unified framework has been proposed aspiring to transform today's Internet into a shopping place for network services, the basis for communication between endpoints. The framework was kept intentionally as generic as possible in an effort to serve as an ambitious and long-term research programme where different technologies, solutions and algorithms may be tried and evaluated.

To this end, the proposed approach may foster cooperation among providers, since those providers that cooperate and negotiate mutually beneficial agreements will be in a better position to provide better services and to promptly react to customers' requests. Moreover, since it will become easier for users to select providers, they will be more likely to select those providers that offer a better cost-benefit relation, and this will only be possible if they have agreements for feedback and measurements on the performance levels and open their infrastructures to customized services.

References

1. Mikhail Smirnov, "Autonomic Communication: Research Agenda for a New Communication Paradigm", White Paper, Fraunhofer FOKUS, Berlin, Germany, November 2003, http://www.autonomic-communication.org/publications/
2. Lidia Yamamoto, "Automated Negotiation for On-Demand Inter-Domain Performance Monitoring", Proceedings of 2nd International Workshop on Inter-Domain Performance and Simulation (IPS 2004), Budapest, Hungary, March 2004, pp.159-169.
3. Elisa Boschi, Salvatore D'Antonio, Giorgio Ventre, "Inter-domain Communication and Data Exchange", Proceedings of 2nd International Workshop on Inter-Domain Performance and Simulation (IPS 2004), Budapest, Hungary, March 2004, pp.65-72.
4. Bruno Quoitin and Olivier Bonaventure, "A survey of the utilization of the BGP community attribute", Internet Draft draft-quoitin-bgp-comm-survey-00.txt (work in progress), February 2002.

5. Samphel Norden and Jonathan Turner, "Interdomain QoS Routing Algorithms", Washington University, Department of Computer Science Technical Report, WUCS-02-03, 2002.
6. Timothy G. Griffin, Gordon Wilfong, "Analysis of the MED Oscillation Problem in BGP", 10th IEEE International Conference on Network Protocols (ICNP'02), Paris, France, November 2002.
7. E. Crawley, R. Nair, B. Rajagopalan, H. Sandick, "A Framework for QoS-based Routing in the Internet", Section 5: "Interdomain Routing", Internet RFC 2386 (Informational), August 1998.
8. Olivier Bonaventure, "Using BGP to distribute flexible QoS information" Internet Draft draft-bonaventure-bgp-qos-00.txt (work in progress), February, 2001.
9. Li Xiao, King-Shan Lui, Jun Wang, Klara Nahrsted , "QoS Extension to BGP", 10th IEEE International Conference on Network Protocols (ICNP'02), Paris, France, November 2002.
10. Cornel Pampu, Henning Schulzrinne, Xiaoming Fu, Cornelia Kappler: "Design of Technology Independent QoS Signalling Protocol for intra- and interdomain environment", First international workshop on Inter-domain performance and simulation (IPS 2003), Salzburg, Austria, February 2003.
11. N. R. Jennings, P. Faratin, A. R. Lomuscio, S. Parsons, C. Sierra and M. Wooldridge, "Automated negotiation: prospects, methods and challenges", In International Journal of Group Decision and Negotiation, 10(2), pages 199-215, 2001.
12. M. Klein, P. Faratin, H. Sayama, and Y. Bar-Yam, "Protocols for Negotiating Complex Contracts", IEEE Intelligent Systems Journal, Special Issue on Agents and Markets, Volume 18, Number 6, pp. 32-38, 2003.
13. P. Faratin, N. Jennings, P. Buckle and C. Sierra, "Automated Negotiation for Provisioning Virtual Private Networks using FIPA-Compliant Agents", The Fifth International Conference and Exhibition on the Practical Application Of Intelligent Agents And Multi-Agent Technology (PAAM-2000), pp. 185-202, Manchester, UK, 2000.
14. P. Faratin, J. Wroclawski, G. Lee and S. Parsons, "The Personal Router: An Agent for Wireless Access", In Proceedings of the AAAI Fall Symposium on Personal Agents, pp. 13-21, N. Falmouth, Massachusetts, US, 2002.
15. P. Faratin, J. Wroclawski, G. Lee and S. Parsons, "Social User Agents for Dynamic Access to Wireless Networks", Proceedings of the AAAI Spring Symposium on Human Interaction with Autonomous Systems in Complex Environments, Stanford, PA, US, 2003.
16. M. Calisti, B. Faltings, "Distributed constrained agents for allocating service demands in multi-provider networks", Journal of the Italian Operational Research Society, Special Issue on Constraint-Based Problem Solving, volume XXIX, number 91, 2000.
17. M. Calisti, B. Faltings, "Agent-Based Negotiations for Multi-Provider Interactions", Proceedings of 2nd International Symposium on Agent Systems and Applications (ASA 2000), Zurich, Switzerland, September 2000.
18. The Foundation for Intelligent Physical Agents (FIPA), http://www.fipa.org/
19. FIPA Contract Net Interaction Protocol Specification, SC00029H, December 2002, http://www.fipa.org/.
20. FIPA Iterated Contract Net Interaction Protocol Specification, SC00030H, December 2002, http://www.fipa.org/.
21. FIPA Brokering Interaction Protocol Specification, SC00033H, December 2002, http://www.fipa.org/.
22. FIPA Propose Interaction Protocol Specification, SC00036H, December 2002, http://www.fipa.org/.

23. FIPA English Auction Interaction Protocol Specification, XC00031F, August 2001, http://www.fipa.org/.
24. FIPA Dutch Auction Interaction Protocol Specification, XC00032F, August 2001, http://www.fipa.org/.
25. FIPA Quality of Service Ontology Specification, SC00094A, December 2002, http://www.fipa.org/.
26. FIPA Network Management and Provisioning Specification, XC00082B August 2001, http://www.fipa.org/.
27. Nicholas R. Jennings, Abe Mamdani, Timothy J. Norman, Jeremy Pitt, "Agreements Work Plan", FIPA work plan (work in progress) http://www.fipa.org/docs/wps/f-wp-00008/f-wp-00008.html, August 2000.
28. FIPA Board of Directors, "TC Agreement Management Resolutions", http://www.fipa.org/docs/output/f-out-00084/f-out-00084.html, April 2001.
29. Chavez and P. Maes, "Kasbah: An Agent Marketplace for Buying and Selling Goods", In Proceedings of the First International Conference on the Practical Applications of Intelligent Agents and Multiagent Technology (PAAM '96), London, UK, 1996
30. P.R. Wurman, M.P. Wellman, and W.E. Walsh, "The Michigan Internet AuctionBot: A Configurable Auction server for Human and Software Agents", In Proceedings of the Second International Conference on Autonomous Agents (Agents-98), Minneapolis, MN, USA, May '98.
31. R.H. Guttman, A.G. Moukas, and P. Maes, "Agent-mediated Electronic Commerce: A Survey", Knowledge Engineering Review, June 1998.
32. G. Cortese, R. Fiutem, P. Cremonese, S. D'antonio, M. Esposito, S.P. Romano, A. Diaconescu, "Cadenus: creation and deployment of end-user services in premium IP networks", IEEE Communications Magazine, Volume 41, Issue 1, pp.54-60, Jan 2003.
33. Object Management Group: CORBA Components, Revision 3.0, OMG document formal/02-06-65
34. Vicente, J., S. Denazis, et al., "L-interface Building Block APIs", IEEE P1520.3, P1520.3TSIP016, 2001.
35. S. Denazis, S. Karnouskos, T. Suzuki, S. Yoshizawa, "Component-based Execution Environments of Network Elements and a Protocol for their Configuration", IEEE - Transactions on Systems, Man and Cybernetics, Special Issue on Technologies that promote computational intelligence, openness and programmability in networks and Internet services, Autumn 2003 (in press)
36. Galis, S. Denazis, C. Brou, C. Klein (ed), "Programmable Networks for IP Service Deployment", Artech House Books, ISBN 1-58053-745-6, 2004.
37. "Overview FAIN Programmable Network and Management Architecture", FAIN Project Deliverable 14, http://www.ist-fain.org/
38. Biswas, J., et al., "The IEEE P1520 Standards Initiative for Programmable Network Interfaces", IEEE Communications, Special Issue on Programmable Networks, Vol. 36, No 10, October, 1998.
39. IETF ForCES, draft-ietf-forces-framework-04.txt, December 2002, http://www.ietf.org/internet-drafts/draft-ietf-forces-framework-04.txt
40. Yang, L., J. Halpern, R. Gopal, R. Dantu, "ForCES Forwarding Element Functional Model", March 2003.
41. David G. Thaler and Chinya V. Ravishankar, "An Architecture for Inter-Domain Troubleshooting", Journal of Network and Systems Management, 12 (2): 155-189, June 2004.

TurfNet: An Architecture for Dynamically Composable Networks

Stefan Schmid, Lars Eggert, Marcus Brunner, and Jürgen Quittek

NEC Europe Ltd.,
Network Laboratories,
Kurfürstenanlage 36,
69115 Heidelberg, Germany
{schmid, eggert, brunner, quittek}@netlab.nec.de

Abstract. The Internet architecture is based on design principles such as end-to-end addressing and global routeability. It suits relatively static, well-managed and flat network hierarchies. Recent years have shown, however, that the Internet is evolving beyond what the current architecture can support. The Internet architecture struggles to support increasingly conflicting requirements from groups with competing interests, such as network, content and application service providers, or end-users of fixed, mobile and ad hoc access networks. This paper describes a new internetworking architecture, called TurfNet. It provides autonomy for individual network domains, or Turfs, through a novel inter-domain communication mechanism that does not require global network addressing or a common network protocol. By minimizing inter-domain dependencies, TurfNet provides a high degree of independence, which in turn facilitates autonomic communications. Allowing network domains to fully operate in isolation maximizes the scope of autonomic management functions. To accomplish this, TurfNet integrates the emerging concept of dynamic network composition with other recent architectural concepts such as decoupling locators from identifiers and establishing end-to-end communication across heterogeneous domains.

1 Introduction

The Internet has evolved from a small research network to a huge, worldwide information exchange that plays a central role in today's societies. A growing diversity of interests in this global internetwork (*e.g.*, commercial, social, ethnic, governmental, etc.) leads to increasingly conflicting requirements among competing stakeholders. These conflicts create tensions the original Internet architecture struggles to withstand.

As one example of an ongoing "tussle" [1], consider the commercial success of the Internet. It has created a large number of competing service providers that aim to outperform one another in order to increase their profits. The result is an increased willingness to forgo agreed-upon standards that allowed a more cooperatively managed Internet to succeed. This paper argues that despite the remarkable success of the Internet architecture – often attributed to its robust design principles – its underlying

M. Smirnov (Ed.): WAC 2004, LNCS 3457, pp. 94–114, 2005.
© IFIP International Federation for Information Processing 2005

assumptions no longer fully match today's networking requirements. In particular, specialized new types of networks, such as sensor networks, mobile *ad hoc* networks and the widespread deployment of "middleboxes" have begun to stretch the capabilities of the existing architecture. This has prompted research into fundamentally different network architectures, such as FARA [2], Plutarch [3], Triad [4] or IPNL [5].

This paper proposes a new internetworking architecture called *TurfNet*, which addresses the limitations of the Internet architecture by accommodating conflicts of interests among different stakeholders and supporting their diverse interests.

The *TurfNet* architecture focuses on interoperation between otherwise autonomous networks. These autonomous networks are modularized according to the inherent boundaries drawn by the different interests of the stakeholder involved. This paper uses the name *turf* to denote such an autonomous network. The term *turf* has an innate connotation to ownership and responsibility that the *TurfNet* architecture reflects. Other papers introduce different terms for similar concepts, such as *regions* [6] or *contexts* [3]. The concept is also related to the Internet's *Autonomous Systems* (AS).

Isolated, autonomous *TurfNets* dynamically compose into new, larger autonomous *TurfNets* that integrate the original networks. The process of dynamic network composition supports the interconnection of heterogeneous networks, such as mobile and *ad hoc* networks, IPv4 networks or IPv6 networks. Composed "super" networks manage this integration by abstracting potential isolation (*e.g.*, over-lapping address spaces) or heterogeneity (*e.g.*, incompatible network protocols) issues among the constituent subnetworks. One mechanism for supporting this heterogeneity is address and protocol translation, but the architecture supports other, equivalent mechanisms as well.

Backwards compatibility with today's Internet is a crucial requirement for any next-generation internetworking architecture. This was arguably one critical mistake during the design of IPv6. The *TurfNet* architecture maintains compatibility with the current Internet architecture by supporting it as one specific network type, along with 3G mobile networks, *ad hoc* networks or sensor nets.

The first part of this paper motivates this research and discusses the underlying design principles of the architecture. Section 3 then outlines the *TurfNet* architecture and explains how it addresses the "new" needs of today's networking requirements. Section 4 describes the basic end-to-end communication across several layers of composed *TurfNets*. Section 5 then discusses the scalability properties of the *TurfNet* architecture. Finally, the remaining sections of this paper compare and contrast the *TurfNet* architecture against other related work and conclude with an outlook on future work.

2   Design Axioms

This section briefly discusses the basic axioms of the *TurfNet* architecture.

Packet switching. Packet-switched networks increase performance and efficiency by multiplexing bursty traffic from different sources onto the same medium. Furthermore, packet switching provides a simple, generic communication framework that supports many different kinds of data flows and requires little explicitly managed state inside the network.

Separation of identity and location. Today's IP addresses denote both the identity of a node as well as its topological location. Several proposals for splitting the two functionalities exist, and the *TurfNet* architecture will adopt this important new concept with a focus on supporting mobility.

Global namespace. The Internet's *Domain Name System* (DNS) [7] is a global, hierarchical namespace based on *Fully Qualified Domain Names* (FQDNs). Other current naming schemes, such as the ones used for the *Host Identity Protocol* (HIP) [8] or the *Layered Naming Architecture* [9], also make use of globally unique names or identities.

Similar to these approaches, the *TurfNet* architecture globally identifies network entities belonging to different, autonomous *TurfNets*. Consequently, the *TurfNet* architecture could either use FQDNs, HIP identities, or any other global namespace – or even different global namespaces at the same time. For simplicity, the remainder of this paper uses the generic term "name" to refer to arbitrary types of global identifiers.

Flexibility in business models. Networking moves from a few monolithic operators to a scenario where the competing interests of owners, roaming brokers, transit network operators, users, and service providers, among others, must be accommodated and balanced. Consequently, a future internetworking architecture must enable, support and manage new business models and complex value chains.

Autonomous *Turfs*. Braden [10] proposes the meta-architectural principle that different regions of the network should be allowed to differ from each other: "minimize the degree of required global architectural consistency." This paper adopts this principle as a necessary enabler for future businesses and diversity between domains.

Inter-*Turf* control interface. Network control must cross domain boundaries, for example, to support address registration and name lookups across individual *TurfNets*. Such functions require a common, high-level inter-*Turf* control interface to exchange control state and configuration information. This facility must not be tied to a specific network protocol – which could be different within individual *TurfNets* – but rather depend on a common data format.

3   The *TurfNet* Architecture

A *TurfNet* is a completely autonomous network domain. To achieve autonomy, every *TurfNet* encompasses its own, independent network addressing mechanism and all associated control plane functions, such as routing protocols, name-to-address resolution, etc. A common, shared namespace to enable inter-*Turf* communication is the only global requirement (apart from the high-level inter-*Turf* control interface). In contrast to today's Internet architecture, *TurfNets* do not rely on globally shared state or pervasive functionality like a common network protocol, a globally shared address space or a global name service.

Another fundamental design choice that supports autonomy of *TurfNets* is the concept of encapsulation. It allows *TurfNets* to fully hide their internal characteristics, structures and policies. Such a modular network architecture allows individual players with potentially competing interests to interoperate in a controlled and protected manner and thus better suites the new requirements of future network communication.

If a *TurfNet* chooses to hide its internals (*e.g.*, network addresses and protocols), external nodes cannot directly communicate with individual nodes of that *TurfNet* anymore. Communication without knowledge of the peer node's local network address (and protocol) requires new network capabilities. In the *TurfNet* architecture, nodes first have to acquire a *Turf*-local representation in the destination *TurfNet*. In essence, each *TurfNet* maps the remote communication peers into part of its local address space. To other local *TurfNodes*, remote nodes appear to be of the local *Turf*.

3.1 Architecture Overview

Figure 1 shows an abstract view of the proposed *TurfNet* architecture. Its key components are:

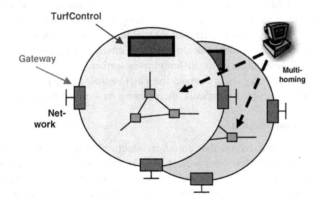

Fig. 1. The *TurfNet* Architecture

TurfControl. The *TurfControl* is a logical, per-*Turf* entity comprised of a *TurfNet's* essential control functions and services. It encompasses all traditional control plane functionalities in the network, such as address allocation, routing and name resolution. It further includes the new *TurfNet* functionality, for example, to manage *TurfNet* composition.

A *TurfNet* handles all its control functionality locally. This is an important prerequisite for maintaining the autonomy of individual *TurfNets*. Because of the importance of the *TurfControl* for the proper operation of a *TurfNet*, it must be resilient. In the case of large (composed) *TurfNets*, distribution and replication of this logical functionality across many nodes will improve scalability as well as resilience.

TurfNode. A *TurfNode* is a network node in a specific *TurfNet*. It communicates with *Turf*-local network protocols and uses local addressing and routing mechanisms.

A *TurfNode* interacts with the local *TurfControl* for all control plane operations, such as address allocation, routing or name resolution.

To support multi-homing as a fundamental part of the *TurfNet* architecture, physical nodes may concurrently participate as full-fledged logical *TurfNodes* in multiple *TurfNets* (see Figure 1). Note that multi-homed nodes do not necessarily act as gateways between the different *TurfNets* (see below).

Gateways. *TurfNet* gateways are special, multi-homed nodes. Besides being part of multiple *Turfs* at the same time, they also actively relay traffic between the different *TurfNets* they are a part of. To enable this functionality, such gateways must be fully operational *TurfNodes* in both *TurfNets* (*i.e.,* they must be able to communicate and have at least one interface in both *TurfNets*).

The main responsibility of these gateway nodes is to relay traffic between the different *TurfNets*. In the case of peering, *TurfNets* use independent network addressing or even different network protocols. Gateways will then also perform the required address and protocol translations. For example, a gateway between IPv4 and IPv6 *TurfNets* will translate between the two network protocols and their respective address spaces. If two *TurfNets* use the same protocols and have compatible addressing, the gateway will simply forward data packets – acting like a traditional Internet router. Whether a *Turf* gateway acts as a traditional router, as network address translator, or even as protocol translator depends on its local environment. Section 4 discusses advantages and disadvantages of the different roles further.

Another task of gateways is to connect the *TurfControls* of the peering *TurfNets*. If *TurfNets* use different control protocols, the gateway must translate control messages.

3.2 Network Composition

The *TurfNet* architecture adopts the central architectural principle of network composition from the Ambient Networks project [11]. In this context, network composition is introduced as a new paradigm that allows co-operative networks to automatically negotiate inter-working agreements through which they establish inter-domain communication. How network composition at the control plane level can facilitate self-organisation is discussed in a related paper [12] by Kappler et al. In the context of this work, on the contrary, network composition is considered a means to inter-connect fully autonomous networks (for example, *TurfNets*), whereby the focus lays on inter-domain communication among heterogeneous networks (including different network protocols and address spaces).

TurfNets can dynamically compose with each other to form new, integrated or interconnected *TurfNets*. Two different variants of this operation are possible, resulting in *horizontal* or *vertical* composition of individual networks.

3.3 Horizontal Composition

When multiple *TurfNets* merge into a single *TurfNet* such that they share a common control plane as a result, they compose *horizontally*. This type of composition fully integrates the original *TurfNets* into the final, merged *TurfNet*. For example, one *TurfNet* could adopt the addressing mechanisms and protocols of the others, or the merging *TurfNets* could all agree on a new set of addressing mechanisms and proto-cols. Figure 2 illustrates this process.

A key characteristic of horizontal composition is that merged *TurfNets* have a sin-gle logical *TurfControl* instance. The original *TurfNets* lose their "identity" after the composition and appear from then on only as part of the identity of the new *TurfNet*. The process of horizontal composition is irreversible – no information about the origi-nal constituents exists that would allow decomposition into the original *TurfNets*.

Despite this loss of identity, horizontally composed networks can still split. This occurs, for example, when a *TurfNet* becomes partitioned due to network link failures. However, partitioning will not typically restore the original *TurfNets* but instead result in an arbitrary set of *TurfNets*.

Finally, note that the *TurfNet* architecture does not prohibit a *TurfNet* from being structured into different administrative domains. These domains, however, are man-agement entities that are not visible in an architectural description of a *TurfNet*.

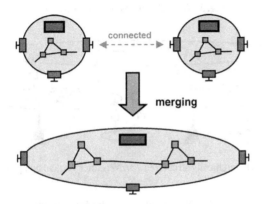

Fig. 2. Horizontal composition of *TurfNets*

3.4 Vertical Composition

Vertical composition, on the other hand, is the process by which *TurfNets* compose such that each individual *TurfNet* preserves its autonomy (with respect to addressing, routing, name resolution, etc.) even after it becomes part of the newly composed *TurfNet*. In this case, two *TurfNets* are said to compose vertically such that one be-comes the governing parent *TurfNet* of the other. Figure 3 illustrates this case and Figure 4 highlights hierarchical structure of this composition variant.

The advantage of vertical network composition is encapsulation of administrative, control and routing functionalities, as well as isolation of internal structures. Because of hierarchical composition, new sub-*TurfNets* may join locally, without requiring global interaction. This reduces the complexity of administrative and control negotia-tions.

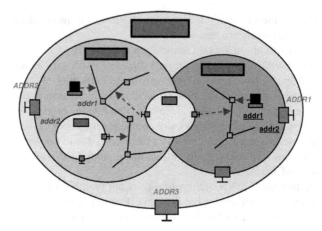

Fig. 3. Vertical composition of *TurfNets*

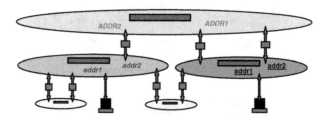

Fig. 4. Hierarchically structured *TurfNets* after vertical composition

Similar to the proposed *TurfNet* architecture, the Internet also contains administrative domains that hide the internal complexity of a domain (*e.g.*, intra-domain routing). However, the main difference to the *TurfNet* architecture is that its vertical composition is not limited to two levels. Furthermore, vertical composition fully separates administrative and control functionalities (*i.e.*, not even the addressing or routing must be globally agreed upon).

Another important advantage of vertical composition is that the transition from today's networking infrastructure requires few changes to existing networks. Specialized gateway nodes facilitate interoperability and integration by translating between different domains. For example, vertical composition can integrate existing IPv4 and IPv6 networks without modifications to existing protocols and protocol stacks.

Communication across vertically composed *TurfNet* boundaries occurs through well-defined gateways, which relay traffic between the different *TurfNets*. If the peering *TurfNets* use the same network protocol and non-overlapping network addresses, a gateway simply forwards packets between the domains, similar to traditional Internet routers. However, if *TurfNets* use different network addressing schemes or different network protocols, a gateway also performs bidirectional network address and protocol translation.

The number of gateways used to compose a *TurfNet* into a parent *TurfNet* depends on reliability and scalability requirements. Large volumes of traffic require a larger number of gateway nodes, to share the processing load of address translation. Multiple gateways also provide resilience in the case of gateway failures.

As illustrated in Figure 4, *TurfNets* can simultaneously compose with several higher-layer *TurfNets*. This is especially important for multi-homed *TurfNets* that peer with several providers.

3.5   Discussion

The previous, brief description of composition approaches has illustrated that horizontal and vertical composition are fundamentally different and address different architectural needs.

Horizontal composition fully integrates two networks into one. In this case, composition only occurs during the initial setup phase of the network integration – afterwards, the composed *TurfNet* operates exactly as a monolithic *TurfNet* would. In other words, composition does not visibly affect performance, scalability, security or other network properties.

Horizontal composition of networks allows integration of networks belonging to the same administration or different administrations. However, horizontally composing networks must all agree on the same address space, address allocation scheme and require a common routing mechanism. One example of horizontal composition occurs between different networks of a single or multiple cooperating network providers. However, composition between a service provider network and its customers' networks requires a different type of composition due to lack of trust and the desire to preserve some level of autonomy between the different parties. Consequently, a more loosely coupled form of composition is needed.

Vertical composition provides this looser form of composition. It enables independent networks with different network architectures that may belong to separate administrations to compose in a way that preserves their individual autonomy and specific internal operation. The gateway nodes, which are configured through the *TurfControls* involved in the composition process, enable the integration and interoperation of the otherwise fully independent networks. The overhead associated with this loose type of composition is acceptable in cases where closer composition is not an option due to administrative concerns, *e.g.*, lack of trust or desire for autonomy).

In terms of the *TurfNet* architecture, the existing Internet topology can be interpreted as consisting of horizontally composed networks, each being its own administrative domain or autonomous system. They share a common address space together with common routing and name resolution functions. The *TurfNet* architecture enables composition of this Internet-wide *TurfNet* with new access networks that feature different control functions, or for example, with vehicle area networks (VANs) that use their own non-IP communication infrastructure, Bluetooth-based personal area networks (PANs), or *ad hoc* networks. All of these networks temporarily or permanently compose with the current Internet through *TurfNet* gateways. The only constraint is that a common, global name space for all of them needs to be place.

Due to space limitations, the remainder of this paper focuses on the – arguably more interesting – vertical *TurfNet* composition, even though both types of composition are equally important pieces of the overall architecture.

4 Basic Operation

End-to-end communication across *TurfNet* boundaries is not trivial due to isolation and heterogeneity of the individual networks. *TurfNet* adopts the decoupling of names and locators from FARA's abstract architectural model [2]. It therefore uses names as global identifiers of *TurfNodes* that are different from the node addresses used for routing.

Because *TurfNode* addresses may not have end-to-end significance (they might merely be transient routing tags for *Turf*-local routing), the architecture uses the name registration and resolution process to find and setup the high-level routing path across composed *TurfNets*. End-to-end communication across *TurfNet* boundaries is thus a product of the following processes: *address allocation, node registration, name resolution* and *packet relaying*.

4.1 Turf-Local Address Allocation

A *TurfNode* joining a *TurfNet* first needs to obtain a *Turf*-local network address using the respective address allocation function of the *TurfControl* (*e.g.*, DHCP [13] or IPv6 auto-configuration [14]). This address only needs to be valid and meaningful within the local *Turf*. How individual *TurfNets* handle address assignment does not affect the inter-*Turf* architecture. Ideally, this process of address allocation should happen in a fully automated way.

4.2 Node Registration

Because individual *TurfNets* may have completely independent network addresses spaces, *TurfNodes* may not be directly addressable from outside their local *Turf* (similar to today's NAT'ed hosts). The lack of a global address space across all *TurfNets* prevents an external node (of another *TurfNet*) from addressing a local node directly. The *TurfNet* architecture exacerbates this problem, because different *TurfNets* may not only have overlapping but also completely different address spaces or network protocols (*e.g.*, IPv4, IPv6, or any other internetworking protocol).

Although NATs are often held responsible for breaking the end-to-end semantics of the Internet – and rightfully so – their hiding capabilities are a central feature of the *TurfNet* architecture. However, *TurfNet* carefully eliminates the disadvantages of NATs by introducing names as explicit, end-to-end node identifiers. Gateways are free to translate addresses and protocols when higher-layer entities bind to static names. Hiding *TurfNet* internals facilitates strong autonomy between *TurfNets* and minimizes shared global state. The ability to hide the internals of networks and the resulting autonomy is becoming critically important in today's commercial Internet, where even companies that own sufficient address space use NATs to control outside visibility of internals of their local networks.

A *TurfNode* that wants to be reachable for nodes outside of its local *Turf* registers its local address with the name resolution service of the local *TurfControl* (step 1 in Figure 5). To achieve reachability from external *TurfNodes*, this registration propagates up the hierarchy of composed *TurfNets* (step 2) via the *TurfControl* of the local *TurfNet*. Note that a *TurfNode* itself cannot register its address in external *TurfNets*, because it is not directly aware of their existence.

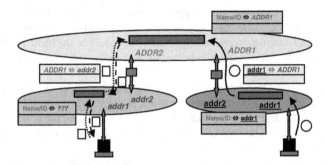

Fig . 5. Node registration and name resolution across *TurfNets* with independent address spaces

If a child and a parent *TurfNet* use separate or even different address spaces, a straightforward propagation of the address registration to the parent *TurfNet* would fail, because the addresses of the registered *TurfNode* have no meaning outside its local *TurfNet*. Consequently, a new *proxy address* for such a node needs to be allocated and registered with the parent *TurfNet*. This process may cause a node to receive different local proxy addresses in each *TurfNet* along the path. The gateways initiate the allocation of proxy addresses. For each address registration propagated to a parent *TurfControl*, the gateway that connects the *TurfNets* requests a new local network address from the address allocation function of the parent *TurfControl*. It then creates the necessary address/protocol translation state between the different addresses. The address allocation function operates in a distributed fashion; for example, each gateway might control its own pool of addresses.

Mappings from names to proxy addresses are *soft state* that times out if not refreshed periodically. This conforms to the requirement of using soft state between *Turf* boundaries. Using soft state for address mappings can also significantly reduce the necessary state in the gateways, because they maintain only the state associated with active nodes at any given time.

4.3   Name Resolution

If a *TurfNode* wants to communicate with a peer node, it requests name resolution through the *Turf*-local name resolution service. If the peer node is part of the same *TurfNet*, this is a fully local operation, as illustrated by steps 3 and 6 in Figure 5.

However, if the peer node is not part of the local *TurfNet*, the local name resolution function propagates the lookup to its parent *TurfNet(s)*, which then try to resolve the name within their respective domains (step 4). Because addresses may only be valid within a single *Turf*, the addresses that name resolution returns may differ from *TurfNet* to *TurfNet*. Thus, if the child and parent *TurfNet* use a different address spaces, gateways must allocate a new proxy address for the resolved name after a successful name resolution by a parent *TurfNet*.

When gateways must allocate a local proxy address (because the child and parent *TurfNets* use different address spaces), they also install the necessary address/protocol translation state for mapping between the different address spaces and/or network protocols. The gateway that receives the successful reply from the parent *TurfNet* can either perform this operation "in-band" or the local *TurfControl* can perform this operation "out-of-band" after it receives the name resolution response.

This address/protocol translation state is also *soft state*, in order to reduce gateway state when only a few inter-*Turf* communications are active. The specific conditions for flushing soft state depend on the individual types of communication. For example, exceeding the maximum idle time of an address translation entry could invalidate the soft state mapping between proxy addresses names.

It is a characteristic of the *TurfNet* architecture that both address registration and name resolution may include address allocation and creation of address/protocol translation state.

4.4 Packet Relaying

End-to-end communication among *TurfNodes* can begin as soon as the address resolution process completes successfully. If both communicating peers belong to the same *TurfNet*, their communication is a completely local process that does not involve inter-*Turf* mechanisms.

If the communicating peers belong to different *TurfNets*, packet relaying involves the following steps.

First, if the peering *TurfNets* use the same network address space and communication protocols, the gateway nodes merely act as traditional routers and forward traffic between the different administrative domains. The *Turf*-local routing protocols then connect (across the *Turf* boundaries) to facilitate inter-*Turf* routing in much the same way as today's inter-domain routing protocols do (*e.g.*, BGP).

Second, if the peering *TurfNets* use independent address spaces and/or different network protocols, *Turf* gateways also perform the necessary address and protocol translations when relaying packets. In this case, the address registration and name resolution procedures have already established the necessary proxy addresses and network address translation state at the gateway nodes along the communication path. Similar to today's Internet NATs, a *TurfNet* gateway adds a dynamic address translation rule for the reverse direction when a local *TurfNode* initiates communication with an external node. For this to work, a *TurfNet* gateway has to maintain network address translation state in both directions and must perform address translation on both source *and* destination addresses. This is often referred to as *twice-NAT* [15].

The result is that communication across *TurfNets* must always follow symmetric paths, because only those gateways have the necessary translation state. This is a change from the current Internet, which supports unidirectional links that cause forward and reverse traffic between two nodes to follow different, asymmetric paths. In reality, however, many Internet protocols and services do not deal with asymmetric paths well, especially if their characteristics (*e.g.*, bandwidth or delay) are sufficiently different. In the future, asymmetric paths across *TurfNets* could be supported by coordinating translation state across different sets of gateways for the forward and reverse path. The management and security aspects of such approaches are currently not well understood and the present *TurfNet* architecture consequently limits itself to supporting symmetric paths only.

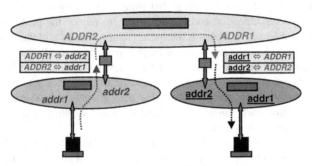

Fig. 6. NAT-based packet relaying in case of peering *TurfNets* with independent address spaces and/or protocols

Figure 6 illustrates packet relaying in detail. Note that although the example shown there is a simple two-stage scenario, the basic mechanism extends recursively to multiple stages. In the left of Figure 6, the initiator of the communication has address **addr1** obtains an external proxy address *ADDR2* in its parent *TurfNet* and creates an entry in the address translation table that maps between those addresses: *ADDR2*⇔**addr1**. Note that different fonts are used for different address spaces: lower-case letters for the lower left, underlined letters for the lower right, and upper-case letters for the top *TurfNet*.

Figure 6 also illustrates the establishment of dynamic address translation state at the *TurfNet* gateways for reverse communication. Each gateway along the path must translate the source address of the sender. For this to work, these gateways must also establish a proxy address for the sender in every *TurfNet* along the transmission path. For the example in Figure 6, this means that the gateways first translate the sender's address (**addr1**) into proxy address *ADDR2* at the top-level *TurfNet* and then into **addr2** at the receiver's *TurfNet*.

When this stage finishes, it has created all required proxy addresses and their corresponding address translation state. Bidirectional communication between the peers is now possible through straightforward address translations.

5   Scalability Considerations

One central assumption of the *TurfNet* architecture is that no hard state exists between individual *TurfNets*; consequently, gateways and name resolution services may only use soft state. Especially critical for scalability and performance is the efficient storage of name-to-address bindings and network address translation state. Because the *TurfNet* architecture decouples domains, each *TurfNet* is free to choose an appropriate solution that satisfies its particular scalability and performance needs.

5.1   Namespaces and Name Resolution

One of the central ideas of the *TurfNet* architecture is the complete autonomy of individual *TurfNets*. They do not require global state, globally unique names or global identities for all communication entities. However, to reduce the problem of global naming to the problem of a global namespace (rather than to a globally distributed name service as in today's Internet), every *TurfNet* provides a local name resolver as a basic functionality of the *TurfControl*. The problem of assigning globally unique names is an orthogonal issue to be solved outside the *TurfNet* architecture.

Local name resolvers in every *TurfNet* that map globally unique names to *Turf*-local addresses allow each *TurfNet* to operate completely autonomous, without the need of external naming services.

In contrast to the Internet's Domain Name Service (DNS), name-to-address mappings in *TurfNet* are soft state, *i.e.*, they need to be updated regularly or they disappear automatically. Therefore, each **(name, address)** tuple has an associated *lifetime*. A *TurfNode* registers its *Turf*-local network addresses with the name service and must then periodically refresh these registrations to prevent them from timing out. Furthermore, **(name, address)** tuples have additional *cache* timers that are similar in function to DNS timers. They indicate how long clients may cache name-to-address resolutions before they must refresh them. A cache timer of zero indicates that clients must perform a new name resolution for each new transport-layer connection. For example, a cache time of zero can be used for mobile nodes that frequently change location.

Figure 5 illustrates inter-*TurfNet* communication, which relies on the fact that local *TurfNodes* that decide to be reachable from remote *TurfNodes* register their location with their parent *TurfNets*. This hierarchical name registration and resolution process ensures that a *TurfNode* that is part of any *TurfNet* in the overall composition can locate any registered node. If the *Turf*-local name service cannot resolve the address itself, the lookup request recursively propagates to the parent *TurfNets* until the name is resolved. Again, this recursive propagation is somewhat similar to the DNS.

A negative side effect of hierarchical name resolution is that top-level *TurfNets* require name-to-address mappings for all registered hosts, *i.e.*, all hosts that choose to be reachable by any node in the composed *TurfNets*. For example, if the Internet were to be rebuilt from composed *TurfNets*, the top-level *TurfNet* would require **(name, address)** tuples for any host that wants to be globally reachable. This example shows the importance of scalability considerations for large (composed) *TurfNets*.

An obvious approach to address the scalability issues is distributing the name-to-address resolution service across many servers. This approach can achieve a high degree of load balancing and fault tolerance. For example, a distributed hash table, such as Chord [16], FPN [17] or Koorde [18], could provide the necessary levels of fault-tolerance and performance for very large numbers of nodes.

The scalability of a name resolver depends entirely on the potential size of its *TurfNet* and the number of nodes that decide to be globally reachable. For example, small *TurfNets* such as an *ad hoc* access network or a Personal Area Networks (PAN) may simply implement a centralized resolver.

The *TurfNet* architecture hides the internal structures and control functions of *TurfNets* and consequently leaves the implementation characteristics of these services to the individual *TurfNets*. For example, whereas a personal-area *TurfNet* may implement the *TurfControl* as a centralized service on a single node, a larger composed *TurfNet* that potentially encompasses a worldwide network with billions of hosts must obviously choose a very different implementation approach to fulfill its specific scalability and performance demands.

5.2   Name Resolution Delegation

This section outlines a particular name resolution mechanism that supports the specific requirements of the *TurfNet* architecture and can scale up to large composed *TurfNets* with billions of *TurfNodes*. The proposed name resolution mechanism is of special interest for the *TurfNet* architecture as it supports highly dynamic address updates despite large-scale deployments (as for example needed for future mobile networks).

The main idea of the proposed mechanism is to split the name resolution process into two steps. The first step simply resolves the *Delegate Address Resolver (DlgAR)*, which is then responsible for the actual name-to-address resolution in a second step (see Figure 7).

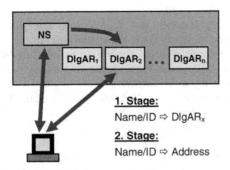

Fig. 7. Delegate name resolution system

Because the mapping from an actual name (or host identity) to the responsible *DlgAR* is expected to be relatively static, the result of this resolution is cacheable for long durations. For example, if this first mapping is based on the name (or host identity) prefix (*e.g.*, all names starting with "a..." map to *DlgAR*$_x$, "*www.ab...*" to *DlgAR*$_y$, etc.), repetition of the first lookup step will be very rarely needed. This achieves a high level of load balancing, because most lookups will only involve the responsible *DlgAR*s.

One benefit of this distribution mechanism is that only a single instance (or at the most a few instances, for reason of fault tolerance) maintains the actual name-to-address mapping[1]. This enables highly dynamic changes, without the overhead of updating many servers. For example, a mobile node that changes its points of network attachment frequently could use this name resolution mechanism to handle the mobility management for new connections.

To further increase scalability of the proposed solution, one could extend the two-stage name lookup mechanism into a multi-stage name lookup mechanism.

5.3   Aggregation of Address Translation State in Gateways

Besides the scalability concerns of name resolution systems in large (composed) *TurfNets*, the architecture also has stringent scalability requirements for gateway nodes. For gateways to allow inter-*TurfNet* communication, they must perform address translations on all inbound and outbound packets. For large (composed) *TurfNets*, this requires the gateways to maintain a dynamic proxy address for any registered or active host that uses it. Besides proxy addresses, the gateway also has to hold the necessary address and/or protocol translation state.

Because a gateway that connects to a top-level (composed) *TurfNet* may provide address translation functionality for potentially huge numbers of nodes, minimizing required state is important. The amount of state required at *Turf* gateways is expected to be of a similar order of magnitude as in today's NAT gateways. Therefore, translation state is not expected to be a limiting factor for scalability.

Nevertheless, state aggregation can minimize state in *TurfNet* gateways. The basic idea is for gateways to allocate dynamic proxy addresses for nodes of sub-*TurfNets* in a way that aggregates addresses, such that one or at the most a few separate entries exist in the address translation table.

The following example illustrates how *TurfNets* can aggregate state. Without loss of generality, the example uses familiar IPv4 addresses. In this example, addresses only have to be unique within a single *TurfNet*. Even sub-*TurfNets* can use overlapping addresses. Figure 8 illustrates state aggregation in the case of an IPv4-based addressing scheme. The top-level *TurfNet* gateway on the left maintains only two aggregated proxy addresses (namely, 10.1.0.0/24 and 10.2.0.0/16) and the relevant address translation state for those sub-*TurfNets*.

This example also shows the advantage of aggregation for the address translation process. In the case of aggregated addresses (for example, the second entry in the address translation table of the top-level gateway: 10.2.0.0/16 ⇔ 10.10.0.0/16), the gateway only has to translate the prefix of the host addresses. In this particular example, only the class B prefix requires translation.

---

[1] Note that today's DNS, which achieves scalability through extensive caching and replication, is not suitable for mobile environments, where dynamic address changes are frequent.

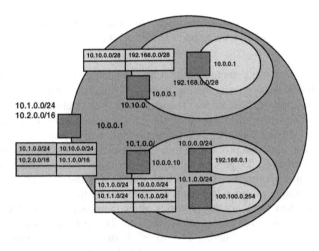

Fig. 8. Aggregation of address translation state

The prefix-based address translation method, based on subnet addresses rather than individual host addresses, has the advantage of only requiring a fraction of the regular address translation state. Prefix-based address translation can also reduce the processing overhead in gateway nodes, as only parts of the addresses must be changed. Particularly in gateway nodes of large composed networks, where address translation state may be highly aggregated, translations would affect only small parts of addresses.

6   Architectural Evaluation

This section evaluates the proposed *TurfNet* architecture through a qualitative examination of the following aspects:

Scalability. The main bottlenecks of the architecture regarding scalability are the explicitly defined gateway nodes that relay inter-*TurfNet* traffic. The relay method – routing, address translation and/or protocol translation – can significantly affect *TurfNet* performance for large numbers of communication hosts and/or data flows. For example, when two core network providers compose their high-speed networks, performing address and/or protocol translation can be problematic. A better approach is to agree on a common address space and routing scheme. On the other hand, when a PAN network dynamically composes with a wireless hotspot network, composition of the different network types through network address translation may be an effective solution.

State aggregation mechanisms can mitigate the problem of state explosion in gateway nodes, as discussed above. However, another approach may be requires to address performance or load problems associated with address translation of all inter-*TurfNet* communication. One obvious approach is to introduce sufficient gateways to load-balance the necessary translation work. Note that the *TurfNet* architecture does not limit the number of gateways per *TurfNet*. Additional gateways will only add some extra control traffic.

Resilience. Inter-*TurfNet* communication relies on dedicated gateway nodes that are able to relay traffic between neighboring *TurfNets*. Hence, the architecture depends on the correct operation of those nodes. This dependency is similar to today's Internet, where NAT'ed networks also depend on the correct operation of the NAT gateway and the Internet at large depends on the correct operation of its routers.

One way to address this problem is to introduce sufficient backup gateways to allow failovers. However, designing a sufficiently fast failover mechanism may prove a challenge. Another approach to improve fault tolerance in the *TurfNet* architecture is through configuration of redundant gateway paths during the initial address registration phase and when resolving the addresses of a communication peer. Note that this could be done by the *TurfControls* along the inter-*Turf* path in a way that is completely transparent to the end nodes. The use of alternative addresses when establishing the end-to-end communication path to a *TurfNode* enables creation of disjoint paths. *TurfControls* along the inter-*Turf* end-to-end communication path recursively choose different gateways and thus disjoint high-level paths for the alternative addresses.

Because communication based on redundant peer addresses flows over a different set of gateway nodes, failures of one gateway on the original path will not affect the others. This allows the communication initiator to switch to a peer's alternative address (thus using alternative paths) in the presence of a gateway failure. The fact that alternative paths through different gateways likely pass through different network service providers further increases resilience, because this approach circumvents problems that could affect whole provider networks.

Performance. The impact of the proposed *TurfNet* architecture on the overall networking performance must be considered carefully. The fact that all inter-*TurfNet* traffic passes through fixed gateway nodes introduces several potential network bottlenecks.

Today's Internet, in some respects, suffers from the same problem, because many home and corporate networks are located behind NAT boxes. In addition, most 2.5/3G mobile access networks relay all external communication through static NAT gateways. Nevertheless, the fact that today's Internet operates – despite the large number of middleboxes – illustrates that performance problems due to NATs are solvable. Correct provisioning of the gateway nodes, both in terms of performance and numbers, is important with an increasing volume of inter-*TurfNet* communication.

One way to address potential performance problems of *TurfNet* gateways due to extensive address translations is dedicated hardware support. Existing hardware solutions for high-speed "label switching" systems (*e.g.*, ATM, MPLS) could in the future also support fast address lookups and rewrites of source and destination addresses.

Flexibility. The key design objectives of the *TurfNet* architecture are creating administratively independent, autonomous networks domains and allowing their dynamic composition. *TurfNets* are fully self-contained and autonomous, even down to the type of addressing and/or the routing protocols they can use. This provides great flexibility for integration and composability of *TurfNets*. Clear administrative boundaries with minimal, but well-defined, control interfaces provide the basis for flexibility.

Mobility. Mobility support is another important criterion for network architectures, because more Internet nodes and subnetworks are expected to become mobile in the future.

Due to the information-hiding capabilities of *TurfNet*, many *TurfNodes* may not be directly addressable. In this event, correspondent nodes will have to address the mobile node through its external proxy addresses. Because this proxy address may not change when the mobile device moves between different *TurfNets*, mobility management can be a local operation that is transparent to the correspondents. In case of a *Turf*-local handoff, the mobile node has to merely inform the local NAT gateway about its new internal address. The hierarchical structure of composed *TurfNets* allows such "local" handoffs at any level in the hierarchy. For example, if a mobile node moves between *TurfNets* that have a common parent *TurfNet*, the handoff only affects the parent *TurfNet*. This local handoff at the parent-level is completely transparent to those correspondent nodes that are located above its parent *TurfNet* or in any other sub-branch of the *TurfNet* hierarchy. Only correspondent nodes in the same branch of the hierarchy are affected by the move. To "repair" inter-*Turf* communication between those correspondent nodes and the mobile node after a handoff, the parent *TurfNet* (where the change of inter-*Turf* routing to the mobile takes pace) signals the respective gateways along the path to update the relevant relaying state. Note that the full specification of *TurfNet* mobility management procedure is currently still under investigation and therefore not yet included here.

7 Related Work

This section discusses related work that also aims at resolving problems of today's Internet architecture.

TRIAD [4] is a recently proposed Internet architecture that tries to resolve the lack of end-to-end connectivity of today's NAT'ed networks by means of an explicit content layer. Similar to the *TurfNet* architecture, TRIAD uses name identifiers rather than addresses for node identification and routing. Because network addresses in both architectures have no end-to-end significance (they are merely used as transient routing tags), both approaches rely on name lookup mechanisms to find and setup the high-level routing path across the independent network domains. However, the main difference between TRIAD and *TurfNet* lies in the way they handle high-level routing. Whereas TRIAD uses source-routing to forward packets, *TurfNet* uses the name registration and lookup mechanisms to configure high-level routing paths and their necessary address/protocol translation state. Another major difference is that TRIAD fully relies on IPv4 support in all transient network domains, whereas *TurfNet* can mask diverse addressing schemes and network protocols in transient network domains through the concept of proxy addresses and protocol translation capabilities at the gateway nodes.

Plutarch [3] is another internetworking architecture that aims to subsume existing architectures like the Internet. Similar to *TurfNet*, the aim of Plutarch is to make the heterogeneity of existing and future networks explicit. To translate communication among heterogeneous network environments (*contexts*), Plutarch introduces the concepts of *interstitial functions*, which allow data to pass between two adjoining contexts. Nevertheless, Plutarch differs fundamentally from the *TurfNet* architecture with respect to naming and routing. Firstly, it assumes different namespaces per context, and secondly, routing is based on sender selection of a context chain and the configuration of the required interstitial functions in gateway nodes along the context chain.

The NAT-extended IP architecture IPNL [5] is another closely related approach. Like the *TurfNet* architecture, it also aims at truly isolating administratively independent IP subnetworks and domains by providing a mechanism to loosely integrate them. The proposed idea also uses NAT middleboxes to integrate networks with a potentially overlapping address space in a way that does not require renumbering. Two fundamental differences to the *TurfNet* architecture exist. First, it does not limit the number of hierarchical composition steps, whereas IPNL considers at the most two levels: NAT'ed *private realms* local networks and global *middle realms* networks. Second, the *TurfNet* architecture does not depend on specific addressing schemes and network protocols, but rather tries to provide a general solution that can integrate many approaches.

A technique similar to IPNL is proposed in 4+4 [19]. Here, address translation occurs also between private and public realms, although in this case it is envisaged to support several "middle" realms. In comparison to IPNL, 4+4 is simpler and allows incremental deployment in today's networks. A fundamental difference between the *TurfNet* architecture and IPNL and 4+4 is that both architectures rely on globally unique host addresses, which consists of the concatenation of the nodes public/global and private/local addresses.

Another related work is the Address Virtualization Enabling Service (AVES) [20]. The key idea here is to virtualize non-IP hosts or hosts that are not globally routable through so called waypoints. The waypoints then act as relays between standard IP hosts and those typically not addressable/reachable hosts. In that sense, AVES is very specific as it only tries to provide bi-directional connectivity for individual hosts. The real overlap between *TurfNet* and AVES lies in the way the waypoint relays are selected. Similar to *TurfNet*, non-IP hosts are dynamically bound to waypoints during the name resolution in a *connection-initiator-specific* fashion.

The concept of 'network pointers' proposed for SelNet [21] is another related approach. Instead of using standard network addresses within data packets, SelNet introduces so called *selectors*, which allow the network pointers (packet handlers) to change the processing semantics of packets as they traverse the network. As a result, SelNet requires a specialized routing protocol that allows mapping of routing information onto selectors.

This section has shown that *TurfNet* is in many aspects related to recent architectural proposals. However, the ability to compose completely diverse autonomous networks and the resultant benefits are a fundamentally new feature of the *TurfNet* architecture. The fact that *TurfNets* can completely mask fundamental difference within individual network domains is especially important in the light of growing tussles in cyberspace [1].

8 Conclusion and Future Work

This paper introduces the *TurfNet* architecture for global, packet-switched internet-works. The architecture addresses the challenges of deploying networks in competitive environments that require means for autonomous control and information hiding.

The *TurfNet* architecture supports horizontal and vertical composition. Vertical composition preserves full autonomy of composed *TurfNets* and supports integration of heterogeneous packet-based networks that use different, non-compatible network-level protocols.

The use of a high-level control interface that only requires a common data format across all autonomous network domains allows control of border gateways that perform the required protocol and address translation for communication across *TurfNet* boundaries.

The autonomy and flexibility provided by *TurfNets* requires the use of soft state. The paper outlines methods that assure a high scalability of the architecture. An important aspect of the *TurfNet* architecture is the high-level or inter-*Turf* routing. This paper discussed the feasibility and realization of inter-*Turf* communication (*i.e.*, packet processing and forwarding), but did not yet specify a specific routing mechanism. Consequently, one area of future work lies in inter-*Turf* routing mechanisms that account for dynamic composition of (moving) networks, as well as performance and reliability for individual end-to-end communication.

Another aspect of the *TurfNet* architecture that requires further consideration is the introduction of virtual "overlay" *TurfNets*. Such virtual *Turfs* can integrate service-specific functionality into the network without complicating its basic functionality.

Finally, another focus of future work lies in evaluating the *TurfNet* architecture through simulations of its scalability properties; especially for large, composed networks. One important aspect of this work is the implementation and measurements of specific prefix-based address translation mechanisms.

Acknowledgements

This document is a byproduct of the *Ambient Networks* project, partially funded by the European Commission under its *Sixth Framework Programme*. It is provided "as is" and without any express or implied warranties, including, without limitation, the implied warranties of fitness for a particular purpose. The views and conclusions contained herein are those of the authors and should not be interpreted as necessarily representing the official policies or endorsements, either expressed or implied, of the *Ambient Networks* project or the European Commission.

References

1. D. Clark, J. Wroclawski, K. R. Sollins and R. Braden. Tussle in Cyberspace: Defining Tomorrow's Internet. Proc. ACM SIGCOMM 2002, Pittsburgh, August 2002, pp. 347-356.
2. D. Clark, R. Braden, A. Falk and V. Pingali. FARA: Reorganizing the Addressing Architecture. Proc. ACM SIGCOMM Workshop on Future Directions in Network Architecture, Germany, August 2003, pp. 313-321.

3. J. Crowcroft, S. Hand, R. Mortier, T. Roscoe and A. Warfield. Plutarch: An Argument for Network Pluralism. Proc. ACM SIGCOMM Workshop on Future Directions in Network Architecture, Germany, August 2003, pp. 258-266.
4. D. R. Cheriton and M. Gritter. TRIAD: A Scalable Deployable NAT-based Internet Architecture. Stanford Computer Science Technical Report, January 2000.
5. P. Francis and R. Gummadi. IPNL: A NAT-Extended Internet Architecture. Proc. ACM SIGCOMM, San Diego, CA, USA, August 2001, pp.69-80.
6. K. R. Sollins. Designing for Scale and Differentiation. Proc. ACM SIGCOMM Workshop on Future Directions in Network Architecture, Germany, August 2003, pp. 267-276.
7. P. Mockapetris. Domain Names - Concepts and Facilities. RFC 1034, November 1987.
8. R. Moskowitz and P. Nikander. Host Identity Protocol Architecture. Work in Progress (draft-moskowitz-hip-arch-06.txt), June 2004.
9. H. Balakrishnan, K. Lakshminarayanan, S. Ratnasamy, S. Shenker, I. Stoica and M. Walfish. A Layered Naming Architecture for the Internet. To appear Proc. ACM SIGCOMM, Portland, OR, USA, August 2004.
10. R. Braden, D. Clark, S. Shenker and J. Wroclawski. Developing a Next-Generation Internet Architecture, July 2000. Whitepaper, available at http://www.isi.edu/newarch/ DOCUMENTS/WhitePaper.ps.
11. N. Niebert, A. Schieder, H. Abramowicz, G. Malmgren, J. Sachs, U. Horn, C. Prehofer and H. Karl. Ambient Networks - An Architecture for Communication Networks Beyond 3G. IEEE Wireless Communications, Vol. 11, No. 2, April 2004, pp.14-22.
12. C. Kappler, P. Mendes, C. Prehofer, P. Pöyhönen and D. Zhou. A Framework for Self-organized Network Composition. Proc. 1st International Workshop on Autonomic Communication (WAC 2004), Berlin, Germany, October 2004.
13. R. Droms. Dynamic Host Configuration Protocol. RFC 2131, March 1997.
14. S. Deering and R. Hinden. Internet Protocol, Version 6 (IPv6) Specification. RFC 2460, December 1998.
15. P. Srisuresh and M. Holdrege. IP Network Address Translator (NAT) Terminology and Considerations. RFC 2663, August 1999.
16. I. Stoica_, R. Morris, D. Karger, M. Frans Kaashoek, H. Balakrishnan. Chord: A Scalable Peer-to-peer Lookup Service for Internet Applications. IEEE/ACM Transactions on Networking, Vol. 11, No. 1, February 2003, pp. 17-32.
17. C. Dubnicki, C. Ungureanu and W. Kilian. FPN: A Distributed Hash Table for Commercial Applications. Proc. HPDC-13, Honolulu, HI, USA, June 2004.
18. M. F. Kaashoek and D. R. Karger. Koorde: A simple degree-optimal distributed hash table. Proc. 2nd Workshop on Peer-to-Peer Systems, Berkeley, CA, February 2003, pp. 98-107.
19. Z. Turanyi, A. Valko and A. Campbell. 4+4: An Architecture for Evolving the Internet Address Space Back Towards Transparency. ACM SIGCOMM Computer Communication Review, Vol. 33, No 5, October 2003, pp 43-54.
20. T.S.E. Ng, I. Stoica and H. Zhang. A Waypoint Service Approach to Connect Heterogeneous Internet Address Spaces. In Proc. of USENIX Annual Technical Conference, Boston, MA, USA, June 2001, pp. 319-332.
21. Christian Tschudin and Richard Gold. Network Pointers. Proc. 1st Workshop on Hot Topics in Networks (HotNets-I), Princeton, New Jersey, October 2002.

A Systems Architecture for Sensor Networks Based on Hardware/Software Co-design

Andy Nisbet[1] and Simon Dobson[2]

[1] Manchester Metropolitan University, Manchester UK
`a.nisbet@mmu.ac.uk`
[2] Department of Computer Science, University College, Dublin IE
`simon.dobson@ucd.ie`

Abstract. We describe the motivation and design of a novel embedded systems architecture for large networks of small devices, tha canonical example being wireless sensor networks. The architecture differs from previous work in being based explicitly on a hardware/software co-design approach centred around the deployment of novel programming language constructs directly onto hardware in order to improve optimisation and expressibility. The programming interface enables the dynamic download and execution of domain-specific code to facilitate the development of context aware pervasive computing systems whose behaviour must adapt to their changing environment. To this end, the architecture implements a virtual machine operating environment based on Scheme and μClinux that encapsulates a CPU core, digital logic, generic I/O, network interfaces and domain-specific programming language composition.

1 Introduction

A sensor network can be viewed as a large-scale distributed system composed of diverse non-uniform hardware devices having both real-time performance and low-power design constraints. Applications running on such platforms must generally adapt their behaviour in response to user tasks, sensed information, dynamic changes in connection topology and temporary/permanent problems with the nodes and communications links present in the network. The adaption can range from simple adjustments of parameters through to partial or complete reprogramming of individual nodes (or indeed the entire network). Thus a sensor network can be viewed as a particularly demanding canonical example of a *context aware pervasive system* where context represents the dynamically-changing distributed environment from the point of view of nodes in the sensor network that are collectively executing one or more distributed applications.

Conventional techniques for the development of pervasive systems have focused either on event-based systems where behaviour is specified using processing tied to events, or on model-based systems using rules applied to a shared context model. [4] argues that both approaches suffer from fragmented application logic, and interactions between rules or events and processing must be analysed

M. Smirnov (Ed.): WAC 2004, LNCS 3457, pp. 115–126, 2005.

in conjunction with environment state information to determine if they result in correct and stable behaviour. *Stability in the face of adaptation* is thus the major design challenge.

Emerging approaches to the development of pervasive systems utilise virtual machines and/or domain specific programming. Maté[2] is one such approach that has developed a virtual machine (VM) for nodes built directly on top of TinyOS[3]. We contend, however, that there may be considerable advantage in applying more advanced programming language approaches directly to sensor networks. Specifically we believe that allowing sensors to be programmed using their own domain-specific language constructs, taking advantage of innovations such as aspect-oriented programming and proof-carrying code, may make a major contribution towards the development of a stable, extensible, comprehensible context-aware systems consisting of thousands of elements.

The vision described in this paper is thus motivated both by developments in sensor hardware and platforms and by recent research in the semantics and construction of programming languages for context aware pervasive computing systems. We seek to combine the notion of context and dynamic domain specific languages into a single infrastructure, by providing:

- a single logical target architecture that can be applied to all nodes in a sensor network;
- an experimental hardware infrastructure based on μClinux[6]; and
- a reconfigurable programming platform using a Lisp- or Scheme[7]-like VM.

Our goal is to investigate language constructs for sensor networks while satisfying real-time performance and low-power design constraints. The dynamic domain-specific aspect we advocate differs from previous work in being based explicitly on a hardware/software co-design approach supporting the deployment of novel programming language constructs directly onto the hardware in order to improve optimisation and expressibility. This is significantly more extensible and portable than (for example) an implementation of Maté extended to dynamically load binary code.

Section 2 presents some basic requirements for hardware in pervasive computing systems in general and wireless sensor networks in particular, and then discusses dynamic domain specific languages from a pervasive systems perspective. Based on this, section 3 offers a research agenda for co-designed context-aware solutions, whose sensor network context is made concrete in section 4. Finally, section 5 concludes with some pointers to the future.

2   Requirements for Truly Pervasive Computing

The development of a pervasive computing application has two logical focal points to its development: the local focus of a node and the collective focus of the network in achieving the objectives of the network application when environment and objectives are dynamically varying. Both focal points have hardware and software components that need to function synergistically, and so are perhaps best treated together.

2.1   Hardware Requirements

Pervasive hardware suffers from a number of design constraints simply by virtue of being targeted at inconspicuous placement in the environment. Sensor networks highlight these constraints particularly clearly – although it is important to realise that they also apply more generally to systems that (for example) include handheld and other elements. A non-exhaustive list of requirements for such networks would include:

Self-organisation and adaptation. A process must discover the availability and quality of network routes that change dynamically with environmental factors, mobility of nodes and temporary and permanent failures of nodes and communication channels. Desirable adaptation features include customisation of the communications protocol, medium access control and routing information. Adaptation is essential as the local and collective roles of the network and how data is processed and communicated are likely to be modified in response to changes in the environment, network applications in execution and tasks requested of applications.

Security mechanisms. Unauthorised modification of network applications (especially for sensor networks) must be prevented. In many applications there may also be stronger privacy guarantees on the ability of outsiders to observe the data sensed or the population of the network.

Discovery. Many networks are self-discovering and self-configuring, in the sense that there is no *a priori* communications or naming topology associated with the elements. The population of nodes can change dynamically over time[1], and applications must be able to tolerate (and preferably benefit from) this dynamism.

Power-aware. Frequently there is no power distribution network physically connected to nodes and power is delivered using batteries and/or is scavenged from energy sources such as light, vibration, movement, stress or fluctuating magnetic fields. A key requirement is the ability to start and stop hardware services and to enter standby modes in order to reduce power consumption. This is of particular importance for any radio interfaces for network communication.

Synchronisation. It is important to be able to synchronize time with groups of nodes, both for applications having fine-grained temporal context and to minimise power by ensuring that all nodes involved in communication during a particular finite time period have powered up and started their radio interfaces.

---

[1] This is even true in augmented materials where elements are embedded at fabrication time. Element and communication failure make such materials dynamic, and it is often too complicated to pre-determine node locations and connections even given that they are embedded in a solid substrate.

Each of these requirements consists of a hardware *and* a software component, with the latter itself consisting of knowledge representation and processing components. Power-aware systems, for example, have the following components:

Hardware. The ability to logically start, stop, suspend and resume components, often in response to events.

Knowledge. A model of the current context of operation and the set of active tasks in order to support decision-making and processing.

Processing. Logic to decide which components may be stopped or suspended in a given situation.

The two software components might typically be fused together, but there are advantages to considering knowledge representation separately from processing. Equally there are co-design challenges in ensuring that the hardware provides the necessary features for software control, and that the software uses these features as well as possible.

2.2   Dynamic Domain-Specific Languages

The evolution of a particular domain-specific programming language can be viewed as the search for the most appropriate mechanisms that express solutions to problems encountered by application developers in the domain. No one language can optimally represent all ways in which to solve a problem: consequently many different languages and techniques have evolved to address different application domains.

The significance of domain-specific languages is that they allow programmers to express directly the concerns of importantance to that domain. By making the concerns explicit, domain-specific languages can provide more structured information to compilers and other tools to inform optimisation.

One approach at unifying disparate languages is aspect-oriented programming[8] where a single language (or occasionally multiple languages[9]) is used to develop separately the individual concerns of a problem. The number and type of aspects are typically fixed at design time: they are then developed and tested separately prior to "weaving" the aspects together late in the development cycle. While aspect-oriented programming has had some successes, it cannot easily integrate new aspects dynamically into the language or program.

A complementary approach is to allow *languages themselves* to be constructed from smaller elements, allowing the construction of domain-specific systems via the composition of language component specifications[10]. A specification might include abstract syntax, concrete syntax, type rules, rewrite rules and perhaps supporting libraries. Libraries need implementations, but the other elements can be specified declaratively. A program can refer explicitly to the language components in which it was developed as part of its source code. A language is then defined by its components and associated evaluators[2] necessary for each

---

[2] An evaluator can be an interpreter or a compiler but we will largely use the term compiler in the text.

of the components. An evaluator for a particular domain specific language is dynamically constructed by compiling the evaluators for its components[5]. A client that downloads a program implicitly discovers the programming language at load time and need only create an evaluator for that particular language as and when execution of the program is required.

Both techniques rely on combining a collection of largely independent fragments in order to create a final program or language. To be useful, there must be both an identifiable set of fragments and a collection of implementations to provide a space within which different combinations may be tried.

3 Characteristic Contributions from Co-design to the Research Agenda for Context-Aware Platforms

Pervasive and context-aware computing rely on the ability to "inject" sensing and computational intelligence into the wider environment, and so encourages the use of microsensing, *ad hoc* wireless networking and advanced reasoning techniques. There is an obvious tension between the requirements: it is relatively easy to construct microsensors and deploy them in a wireless network, but their small size and low power mitigates against including many of the advanced software techniques that are otherwise highly appropriate for managing the network and its results.

A pervasive computing network thus presents a programming platform operating under a unique set of constraints (section 2). It seems unlikely that programming languages evolved for different environments – desktops, servers, and even relatively high-power stand-alone handheld devices – will capture these constraints effectively. This is important both for systems designers (who may not get the best out of their systems) and developers (who will struggle with an inappropriate conceptual model and mode of expression).

However, the most important consideration comes from the ability to deploy sophisticated software anywhere in the environment. Pervasive computing in handicapped by being *asymmetrical*: most of the processing power resides in large dedicated computers. A good example is when assets are tagged with RF-ID tags: the infrastructure (typically a building) can "see" the asset tag, but the asset cannot respond to or make its own determinations about its environment. Constructing applications from networks of low-power nodes goes some way to restoring symmetry to the situation, in that they allow computing and sensing *within* (rather than simply *of*) the asset base. This is also important for scalability.

We contend that the way to address these issues is **to allow the language used in developing pervasive applications to be designed alongside the sensing infrastructure.** This does not preclude familiar constructs or re-use of ideas from other domains – actually quite the contrary – but suggests that **some novel forms will contribute strongly to the effective use** of such networks.

This tight coupling suggests that co-designed network elements may make some distinctive contributions to the research agenda in context-aware, self-deploying and self-managing pervasive computing. Without ignoring the other issues of connectivity, protocols, security, discovery and so forth, here we concentrate on these novel contributions.

Desktop and server systems, despite differences in operating system and programming language, present an overwhelmingly homogeneous platform for developers – a considerable effort has been expended to make this so. Embedded networks are far more heterogeneous, and there is a danger that software will become too targeted at individual elements' capabilities to be able to deal with failure or relocation. We need to understand **what are the correct abstractions for targeting with heterogeneous networks efficiently without over-commitment** to particular details.

Pervasive – and especially sensor – networks can involve large numbers of elements (of the order of thousands). This is far larger than anything dealt with by all but a few distributed systems projects. We have little understanding of **how to efficiently distribute fine-grained functionality** in such systems. Such knowledge as does exist (largely from the high-performance computing communities) deals with static situations: **how we retain performance (in the widest sense) from fine-grained systems that need to constantly re-configure?** It is important to remember that "performance" needs to be understood broadly, as many pervasive computing systems may (for example) stress low power consumption at the expense of processing capacity.

Domain-specific languages suffer to some extent from an over-abundance of flexibility: developers require *some* stability, and one might argue that even a sub-optimal stable core is preferable to a system that is technically better but moving too fast to become expert at. Not all network and sensor features *need* language features: the question therefore arises as to **what is the correct methodology for determining the correct contributions of hardware and software within the co-design?**

It is also easy to forget that computers almost never run a single application, and this is unlikely to change for pervasive computing systems. The network will run several "applications" simultaneously, for different users and with differing (and possibly conflicting) abstractions of the outside world. **How should pervasive computing applications co-exist?** – both at the multiple semantic level as discussed in [4] and at the more prosaic level of allowing different applications, and possibly using completely different programming abstractions and languages. This is an attractive motivator for re-configuring the software capabilities of the network while retaining continuous service.

4   The Sensor Networks Perspective

To make the above agenda more concrete, for the remainder of this paper we will focus on deploying our ideas in the context of wireless sensor networks. Our detailed architectural choices are conditioned by balancing the desire for an open, simple, extensible, rapid development platform against the desire for a solution that can be tested in realistic environments when appropriate. This has led us to choose a hardware architecture based on Field-Programmable Gate Array (FPGA) technology, coupled with a highly dynamic software platform.

4.1 Hardware

The target architecture is based loosely on figure 1 and consists of power sources, FLASH, SRAM and possibly DRAM memory, CPU core, general purpose I/O, RF communication unit and digital logic with μClinux as the embedded operating system. We intend to prototype the architecture using Xilinx FPGA technology to implement the CPU core, digital logic and general purpose I/O. The CPU core is currently planned to use the OR1K open core because a stable μClinux port exists and the core has been successfully synthesised both in silicon (by Flextronics) and in FPGA devices. Although it may not be the most appropriate core for low-power sensor nodes, in theory the core can be used in all node classes in sensor network architectures. Techniques such as clock gating can be used to dynamically switch the processor and other functional units to low-power standby modes[3].

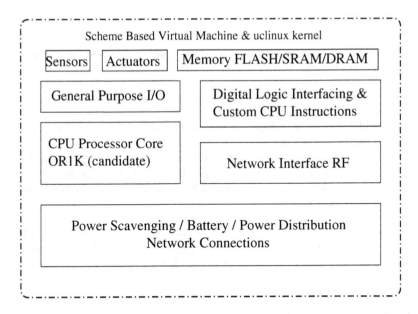

Fig. 1. Proposed target architecture

---

[3] If the sensor node requires ultra low power operation in an energy scavenging environment then it is necessary to use aggressive techniques such as asynchronous logic to implement the architecture as a custom mixed-signal ASIC device.

The digital logic of the FPGA device can be used to interface to general purpose I/O, this will be necessary to connect sensors, actuators, memory and network interfaces (both RF and conventional) to the CPU. Obviously a physical digital interface to the RF will need to be presented to the general purpose I/O. Custom medium access controls can be implemented entirely in digital logic or with some software assistance. FLASH memory can be used to store FPGA configurations and boot images of uClinux and the Scheme based virtual machine operating system infrastructure. The digital logic can also be used to implement custom hardware accelerated user-instructions for the CPU core. Processing that does not map well to the CPU instruction set, or whose computationally requirements make it difficult to meet real-time performance constraints are candidates for implementation in hardware logic. Nodes requiring digital signal processing of audio/video data may require such functionality. The ability to add domain specific processing units in flexible digital hardware means that the target architecture should be capable of offering the performance necessary to implement high bandwidth sensors and gateway nodes. Additionally the use of our proposed architecture makes it possible to implement a systematic method for adding new sensing/actuating hardware that is accessible to our Scheme based software programming platform.

4.2   Software

We have chosen to base our programming platform on the Scheme language[7], for a number of reasons:

1. Scheme has clean semantics and concise syntax that can easily be supported on an embedded system;
2. it provides a rapid prototyping environment for sensor networks that can be easily simulated on desktop computers; and
3. it provides a scripting-based interface to programming sensor networks (as advocated as a useful feature in [11]) that will reduce the level of hardware knowledge required by users, without compromising the possibility of compilation and analysis.

We rejected the Java-based solution of [10] as too heavyweight for a large (and growing) number of sensor network applications, for which supporting a Java VM is either inappropriate or impractical. We rejected a C-based solution for reasons of complexity for application programmers.

4.3   Combination: μClinux and Scheme

A number of attempts[2, 13, 14] have been made to use high-level scripting languages and interpreters in order to simplify application development and to maintain code portability without sacrificing precise control over hardware. The problem then becomes how to maintain a high degree of efficiency alongside virtual access to hardware resources. We have chosen Scheme as the basis for our scripting language (Common Lisp would also have been a valid choice, although more complex). The key question is: how do the Scheme virtual machine (VM) and μClinux interact? This determines to what degree we virtualise access to hardware and the overhead in using this abstraction.

The design options are for Scheme to be positioned:

1. Directly on top of the bare hardware. This would make it possible to construct the entire operating system in Scheme, and μClinux would not be required. Whilst this would be an interesting research direction it is likely to severly limit the space of designs and concepts that might be required as in Movitz[1] where Common Lisp was considered.
2. As a conventional program in a process like the shell. This would be similar in nature to running Scheme in a terminal window on a desktop computer. This approach is by far the easiest to implement but it is the least flexible in terms of operating system customisation. A suitable example of this is the Scheme shell scsh [15] which provides a scheme based scripting language and a posix interface with both high and low-level networking support.
3. Similar to the approach of Movitz, where Common Lisp is used to provide a framework for experimenting with kernel-level development programmed in Lisp. Movitz does not directly support threads and processes, nor does it have an implementation of SLEEP or contain assumptions about how to measure time.

In our work we envisage that the ideal place to position the Scheme VM lies somewhere between 2 and 3 but closer to 3.

There are essentially two choices for implementing a novel language on a sensor network. The traditional approach is to cross-compile the language from a standard desktop host, generating appropriate machine-language instructions. The generated code can be as efficient as handwritten code, although in practice it is typically significantly less so. However, cross-compilers are difficult to develop, debug and optimise.

The alternative is to provide a VM running on the sensor platform itself, accepting the performance and space penalties in order to improve flexibility and ease of development. This is practical only for very small VM run-time systems.

We are exploring both options, but tending towards the latter. Our reasoning is that – paradoxically – there are *fewer* power and space restrictions on a sensor network than in traditional distributed systems because of the sheer number of elements that can be deployed. The challenge is to provide a suitable distribution and co-ordination framework within which to deploy applications over a large number of elements. Using a VM on the elements allows us to focus on this challenge rather than on efficient cross-compilation.

We plan initially to implement 2 in order to provide a working experimental platform at the earliest possible opportunity whilst further researching an appropriate level for Scheme which will enable a sufficient degree of operating system customisation as part of the dynamic domain specific language definition. Clearly our implementation must make modifications to the Scheme virtual machine in order to support the special requirements of sensor networks and dynamic domain specific languages.

4.4   Evaluation Targets

Sensor networks may be used in a variety of real-world scenarios, ranging from earth science and monitoring to security and military applications. Any sensor network architecture must demonstrate an ability to target one or more of these application domains efficiently. The following applications illustrate three broad paradigms with which to evaluate our work, and show how our architecture can improve the ways in which they are addressed.

Habitat/environmental monitoring. Nodes are used to sense features of the habitat that are of interest to scientists and environmental protection agencies over a period of months or even years. The network senses, processes and funnels data towards gateway nodes that are connected to the internet using standard protocols. The data may be pushed or pulled dependent on whether the sensor nodes or tasks (queries) from gateway nodes are the active entities initiating communication. A tree-based routing network must be constructed and maintained. Low-power operation is of prime importance for this application class but fine-grained synchronisation of nodes is usually of low importance. Whether the nodes are fixed or mobile is largely dependent on the habitat, for example sensor nodes circulating in a water system such as a lake will be mobile but nodes deployed by air drop onto a land mass are likely to be fixed.

Shooter localization. The aim of the application is to determine the origin of a bullet or any other projectile in an urban environment. The nodes must sense the shock waves due to the projectile with a high sample rate and fine-grained time synchronisation in order to forward their data onto a gateway node and/or a server where the localisation is normally computed centrally. Power consumption will be significantly higher than in environmental monitoring.

Pursuer-evader/traffic management. The aim of this application is to track the movement of one or more evader robots. The network must route this information to one or more pursuer robots using a routing protocol that exploits knowledge of geographic position information. An obvious extension of this application paradigm would be the more general problem of traffic management where sensors are present in cars, traffic lights and CCTV and speed cameras. The goals and tasks of this traffic application encompass congestion reduction, enforcing road/driving regulations for safety and informing law enforcement and accident and emergency services of appropriate events requiring their intervention.

In each case there are clear hardware and software constraints that must be met by any proposed solution. Our approach is to address these constraints *via* co-design, ensuring that the appropriate language constructs are backed-up by appropriate hardware capabilities. Engineering such solutions pose an interesting challenge: how does one determine the success of a language construct, especially in conjunction with a hardware platform? There is little clear existing engineering methodology to apply to this problem.

5   Conclusion

We have motivated and presented the design of a new architecture for the nodes of a sensor network. The architecture differs from previous work in being based explicitly on a hardware/software co-design approach supporting the deployment of novel programming language constructs directly onto the hardware in order to improve optimisation and expressibility.

Although we have stressed the co-design aspects with reference to small devices, the software techniques can be applied to more traditional platforms as well. This means that a similar domain-specific language could be used across a range of scales, with (for example) some language features being (de)selected on some platforms.

We are currently completing feasibility studies on the components of our proposed architecture, prior to initial development work. Our immediate research challenges are to determine appropriate abstractions for the construction and deployment of the embedded systems architecture from hardware and software perspectives. We intend to evaluate our work against a range of applications, both to check the qualities of individual solutions and to derive methodological understanding that aids the creation of complex co-designed sensor networks.

References

1. Fjeld, F.V.: Movitz: Using Common Lisp for kernel-level programming on commodity hardware. Workshop on Evolution and Reuse of Language Specifications for Domain Specific Languages at ECOOP 2004.
2. Levis, P., Culler, D.: Maté: A tiny virtual machine for sensor networks. In Proceedings of the 10th International Conference on Architectural Support for Programming Languages and Operating Systems (ASPLOS 2002), pages 85-95, San Jose, California, October 2002.
3. Hill, J., Szewczyk, R., Woo, A., Hollar, S., Culler, D., Pister, K.: System architecture directions for network sensors. In Proceedings of the 9th International Conference on Architectural Support for Programming Languages and Operating Systems (ASPLOS 2000), Cambridge, USA, November 2000.
4. Dobson, S., Nixon, P.: More principled design of pervasive computing systems. In Proceedings of Engineering for Human-Computer Interaction and Design, Specification and Verification of Interactive Systems (EHCI-DVIS04), To appear in LNCS.
5. Dobson, S.: Creating programming languages for (and from) the internet. Workshop on Evolution and Reuse of Language Specifications for Domain Specific Languages at ECOOP 2004.
6. uClinux Embedded Linux Microcontroller Project Home Page: http://www.uclinux.org/

7. Lord, T.: Guile Scheme, Technical Report (1996), online 'info' documentation, Free Software Foundation.
8. Kiczales, G., Lamping, J., Menhdhekar, A., Maeda, C., Lopes, C., Loingtier, J.M., Irwin, J.: Aspect-oriented programming. In Akşit, M., Matsuoka, S., eds.: Proceedings of the European Conference on Object-Oriented Programming. Volume 1241 of LNCS. Springer-Verlag (1997) 220–242
9. Lafferty, D., Cahill, V.: Language-independent aspect-oriented programming. In: Proceedings of the ACM Object-Oriented Programming Systems, Languages and Applications Conference (OOPSLA'03), ACM Press (2003)
10. Dobson, S., Nixon, P., Wade, V., Terzis, S., Fuller, J.: Vanilla: an open language framework. In Czarnecki, K., Eisenecker, U., eds.: Generative and component-based software engineering. LNCS. Springer-Verlag (1999)
11. Hill, J. Horton, M., Kling, R., Krishnamurthy: The Platforms Enabling Wireless Sensor Networks. Communications of the ACM, Volume 47, number 6, pages 41-46, June 2004.
12. Levis, P., Madden, S., Gay, D., Polastre, J., Szewczyk, Woo, L., Brewer, E., Culler, D.: The Emergence of Networking Abstractions and Techniques in TinyOS. Proceedings of the First USENIX/ACM Symposium on Networked System Design and Implementation, NSDI 2004.
13. Girod, L., Elson, J., Cerpa, A., Stathopoulus. T., Ramanathan, N, Estrin, D.: Em*: A Software Environment for Developing and Deploying Wireless Sensor Networks, Technical Report 034, Centre for Embedded Network Sensing, UCLA Computer Science Department, Los Angeles, USA, 2003.
14. Madden, S., Szewczyk, R., Franklin, M., Culler, D.: Supporting Aggregate Queries over Ad-Hoc Wireless Sensor Newtworks. Proceedings of the Workshop on Mobile Computing and Systems Applications, New York, USA, June 2002.
15. Scsh - The Scheme Shell Home Page: http://www.scsh.net/

Challenges in Communications Research Beyond the VICOM Project

F. Vatalaro[1], G. Cortese[2], F. Davide[2], A. Detti[1],
M. Leo[3], P. Loreti[3], and G. Riva[4]

[1] Dipartimento di Ingegneria Elettronica - Università di Roma "Tor Vergata",
and CNIT,
Via del Politecnico 1, 00133 Roma, Italy
info@vicom-project.it
[2] Telecom Italia Learning Services S.p.A., Rome, Italy
[3] CNIT, Consorzio Nazionale Interuniversitario delle Telecomunicazioni, Italy
[4] Applied Technology for Neuro-Psychology Lab., Istituto Auxologico Italiano,
Milan, Italy

Abstract. The VICOM (Virtual Immersive COMmunications) project is a
three-year project funded by the Italian Ministry of Instruction University and
Research aiming at investigating innovative communication paradigms. The
project represents a wide coordinated effort focused on integration of immersive and wireless technologies in view of the fourth generation of mobile communications. The main goal of the project consists of the design of a wideband
system architecture for immersive services and of its validation through two
distributed large test-beds. Starting from VICOM ongoing experiences some future challenges and objectives for the future situated and autonomic communications technologies are envisaged in the paper.

1 Introduction

The VICOM (Virtual Immersive COMmunications) project [1] is a project funded by
the Italian Ministry of Instruction University and Research (MIUR) focused on investigating innovative communication paradigms. It is a three-year project started in
November 2002.[1]

The main project goal is the design of a communication system architecture able to
provide mobile immersive services. The architecture effectiveness will be demonstrated in two service test-beds identified in the project as Mobility in Immersive
Environment (MIE) and Virtual Immersive Learning (VIL), respectively. The test-
beds aim at being one first step towards new communication models focused to

---

[1] VICOM partners are: CNIT, a consortium of universities acting as the coordinating partner,
and involving researchers from several universities, the Italian National Research Council
(CNR), through their Bologna and Pisa units, the Polytechnic of Milan, the ISCTI (Istituto
Superiore delle Comunicazioni e delle Tecnologie della Informazione) of the Ministry of
Communications, and Telecom Italia Learning Services.

M. Smirnov (Ed.): WAC 2004, LNCS 3457, pp. 127–138, 2005.

achieve a natural interaction with the communication media. VICOM objectives are integration of the immersive and virtual technologies with wireless access technologies, the development of ubiquitous interaction tools between humans and virtual and physical environments, the seamless integration of immersive interfaces and sensors with the environment to induce perception of a natural interaction, the development of non-invasive multimodal interfaces, wearable devices, "aural" networks and the evaluation of their ergonomics, the degree of user acceptance, as well as the psychological effects.

Starting from the description of the VICOM ongoing research the paper illustrates some future challenges and objectives for the future situated and autonomic communications technologies.

2   VICOM Research Framework

Multimedia mobile communications are starting to face the challenge of integrating audio, video and sensing interfaces, in order to realize new forms of services. Among these, mobile immersive services enriched with virtual contents will play a significant role: this is the technological framework for the VICOM project. In the future of telecommunications, beyond the present so called third generation (3G), virtual and immersive communication services will induce an augmented reality experience in the user through the integration of pervasive communication technologies and virtual multimedia contents. A unified architecture will seamlessly integrate all the needed system and service features. This is one first step towards the development of a auto-reconfigurable, scalable and technology independent communication system foreseen by the situated and autonomic paradigm for the 2020s.

To reach this purpose a system architecture has been designed to transparently support ambient intelligent services, to provide advanced context aware functionalities, to make available user content adaptation services and to offer self-organized network configurations. The system architecture, now under development, aims at providing a first benchmark to experiment virtual and immersive technology components in an exemplary distributed scenario. The selected service scenario is the so-called "VICOM campus" comprising multiple real campuses, augmented by virtual contextual information as dynamically required to service the user. Therefore, the campus will be a mixed reality distributed service area in which the user may benefit, on his/her personal devices and/or on ambient devices, of a predefined set of immersive services, based both on person-to-person and person-to-system communications.

Main research areas are communications, ambient and service intelligence, and virtual multimedia contents delivery. The specific studies undertaken are focused on novel communications paradigms [2],[3]. Nevertheless, the system architecture intends to be effective also with today's off-the-shelf virtual reality and wireless technologies. Moreover the system architecture aims to be "plug & play", seamlessly interfacing different technology components, which will live together in the experimental system. To this aim a set of standard interface are being defined.

Fig. 1 summarizes the research topics providing foundations to the two VICOM immersive technology test beds (MIE and VIL) to be developed.

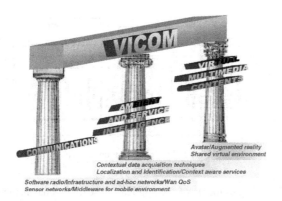

Fig. 1. VICOM research pillars

The MIE test-bed conjugates immersive and mobility aspects, based on wideband wireless techniques, in order to offer "light" virtual multimedia contents to the user on-the-move, both pedestrian (indoor, outdoor) and vehicular. On the other hand, the VIL test-bed is mainly focused on "heavy" virtual contents distribution within a fixed high-speed content delivery network. In both the test beds service intelligence is achieved by means of a context-aware service approach, while the ambient intelligence components are off-the-shelf hardware components (microphones, cameras, etc.) enriched with software applications specially developed or adapted on purpose.

3   Architecture and Services for the Test-Bed

3.1   Architecture

The system architecture is designed to support the main system functionalities: shared context function and content adaptation. The shared context function is in charge of disseminating context data among the users. The content adaptation function changes and/or scales the services presentation accounting for the device capabilities [5]. These functions will generate a data flow composed of natural and synthetic flows, such as avatars or contextual iconic representations that enable several forms of presentation. In fact the data collected in the test-bed environment, gathered by heterogeneous sensors (i.e. cameras, microphones, wireless sensors, etc.), and low level context data (i.e. number of persons in a room, available communications ports, etc.) will be fused together in order to provide the application layer high level context information (i.e. in the room there is a meeting and which communication means are usable).

Moreover a middleware layer will provide both context data dissemination and session-oriented communications; the network functionalities will provide the needed communication capabilities integrating fixed and wireless network elements. It includes ad-hoc networking features offering the best QoS management. Fig. 2 illustrates the solution for the architecture of the software platform.

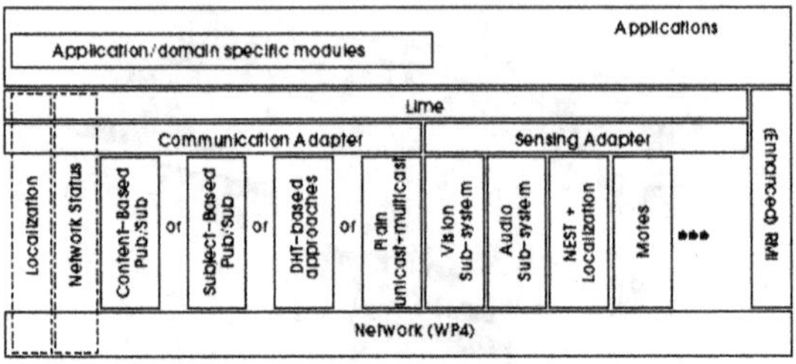

Fig. 2. The VICOM architecture

The analysis of requirements of the application scenarios has made clear a number of requirements for the middleware useful to support the development of the various functionalities that require support for:

- Proactive and reactive operations. Applications must be able to proactively request for information, as well as be asynchronously notified of relevant events.
- Content-based information access. Applications must be able to select the information they need to access, using filtering criteria that are not necessarily fixed a priori and that are based on the information content itself.
- Data sharing. User applications need to share with others their data, both of applicative and contextual nature, and need to do so regardless of possible network reconfigurations.
- Decentralized, peer-to-peer programming model. If the fundamental requirement of mobility is to be addressed radically, the programming model cannot explicitly rely on a server. Even if mobility is not part of the picture, the large-scale characteristics of VICOM scenarios demand for a high degree of decentralization.

To support these functionalities the middleware layer, used initially in static environments and specifically redesigned for the mobile environment, is adopted as the main interface towards the applications. It introduces rules for the management of data in the shared spaces. This programming model encompasses all the four aforementioned requirements.

The software architecture foresees that the middleware layer provides the overall programming interface for the user applications and that the Communication Adapter, below the middleware, decouples it from the specific transport layer. This way allows that the middleware supports flexible and cooperative interactions among user applications, both in ad-hoc and in infrastructure mode, which are specific of the MIE scenario, as well as for the VIL test-bed.

The counterpart of the Communication Adapter is the Sensing Adapter, which effectively decouples the middleware from the specific of the sensing techniques employed to extract contextual information from the environment. The middleware essentially takes care of disseminating context information across the system. The design context information management is central for the effective development of immersive systems architecture.

Fig. 3 shows the basic logical components required for context information processing, i.e. the components for generation of context information from sensors and for distribution of these data to applications.

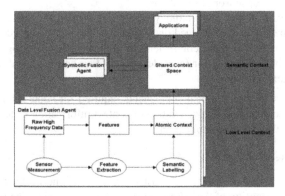

Fig. 3. Context Data Acquisition, Fusion, and Distribution

Low level sensor data are filtered, acquiring raw data and performing some analysis on them, into meaningful 'semantic context' information.

At level of semantic layer the context data are published in a shared data space; applications and symbolic fusion agent components access and possibly subscribe to this information. In this last case, they receive automatic notification of new information. Moreover symbolic fusion agents generate additional context information at a higher level of abstraction.

The user applications access the context space, either in a proactive (query) or publish subscribe style. Applications must have a way to discover sources of context information (i.e. to join the context space) as well as the formats and semantics of data in the space.

- The Shared Context space pattern implements a common, distributed "blackboard" offering publish-subscribe style coordination to a number of peers.
- Providers of information to this space include:
- sensors in the physical environment (cameras, environmental sensors, etc.);
- sensors in the user's Personal Area Network (GPS, motion sensors, cameras, speakers etc);
- inference components for high level;

In addition to basic publish-subscribe mechanisms the context space should provide more sophisticated features such as:

- the ability for applications to obtain time series in addition to single observations;
- aging mechanisms (allowing context information to 'expire' after some predefined time);
- history mechanisms to ease storage of observations, and queries on these observations;
- policy-based control over the quantity of data stored and refresh intervals for sensor observations;
- extensibility and ontology discovery: data in the context space should belong to well-defined ontologies; but flexibility is needed to allow sensor and inference components to implement new ontologies. Middleware should provide for (provider) publication and (client) discovery of context ontologies.

Fig. 3 shows the basic logical components required for context information processing, i.e. the components for generation of context information from sensors and for distribution of these data to applications.

3.2 Test-Beds

For the MIE test-bed, we plan to extensively use ad-hoc networking during the experiment as an extension of the infrastructure wireless coverage. The selected person-to-system trial application is the virtual guide service that augments real world viewed by the user with guide information, in order to create the immersion of the service in the natural user environment. The main functionalities that will be validated with this test-bed are:

- prove the sensing, storing and distribution of the context data (location and identity) both in ad-hoc and in infrastructure wireless environment;
- use of spatial model of the ambience within the application;
- use of augmented reality for the presentation of virtual contents.

In the VIL test-bed, the trial service is the immersive teaching in which each student is reported on a virtual room where real classrooms or home stand alone users are represented by video-stream or by "avatar" depending on their communication capabilities. The teacher is provided with mean to enhance his "student feedback" with respect to actual e-learning approach. Moreover, new interfaces based on virtual reality representation enable the students in accessing remote laboratory instruments in a shared virtual environment fashion. In this test-bed, we resort to fixed network with QoS management, due to the large bandwidth request. The main functionalities that will be validated with this test-bed are:

- prove the use of novel technologies of coding, transport and presentation of natural, synthetic, and mixed multimedia flows enriched with contextual "icons";
- verify the psychological impact of this form of novel e-learning on users.

3.3 User Applications

The virtual immersive guide application will guide the test-bed user towards a specific target. As soon as the user will enter in the test-bed area he will be identified and located. The technologies for the identification and for the localization will be of several different kinds, nevertheless a data fusion module will provide to use the data to extract a single information on the user to disseminate in the shared environment. After the incoming phase the user will choose his target and he will be guided by means of instructions displayed on his device. The instructions and the messages exchanged with the environment and the other devices will be different as function of the user device; they can be provided by a speaking avatar, or by an arrow superimposed on the user view, or by simple textual messages. When the user will reach the target he will be able to get some more information by viewing some information tags that he meets also during his path (Fig. 4).

Fig. 4. Virtual guide examples

With the application for the interaction with the environment the user will be able to interact with the surrounding ambient. Superimposed to the scene he/she is viewing and will able to display some information tags. These tags will be placed both on objects such as printers or computers and on personal devices. For each tag the user will be able to make different actions using a pointing tool. For example by pointing on a printer he will display a menu containing several choices that will be able to print a document command or execute a file transfer with another user (Fig. 5).

Fig. 5. Environmental interaction example

The virtual cooperation in ad-hoc environment application will allow to interconnect many users in an ad-hoc environment. The test-bed scenario will be composed by several independent ad-hoc islands (Mobile Ad-Hoc Network - MANET) interconnected among them. In this environment each user will be able to communicate both directly with the users inside his MANET or the remote users located in external MANET. As shown Fig. 6 a user can join a discussion room talking with users located in one or more different MANETs.

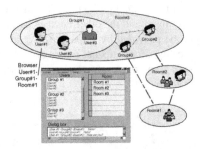

Fig. 6. Cooperation in ad-hoc environment example

In the test-bed VIL the immersive learning application will provide the user of a multimedia board in which natural and synthetic data flows will merge to offer an interactive interface. Using context-data the synthetic objects will superimpose to natural data flows to improve the interaction during one e-learning session. Moreover immersive technologies will increase the sense of being there so the user will have the experience to be in a virtual classroom. The context data will play an important role in this test-bed because the environment data (i.e. temperature, number of people, etc...) will be used to increase the information content of data flows. The virtual classrooms will be equipped with environmental sensors to provide for the identification and the localization services. They will be interconnected among them and the middleware will share the context data detected. The context management engine will be able to recognize human gesture and other conventional signs useful to have a natural interaction among all the users. Main targets will be the capability to perform the identification and to precisely locate the users that take part to the lesson. Another issue of the test-bed will be the capability for the system to adapt the data flows to the connection availability (Fig.7).

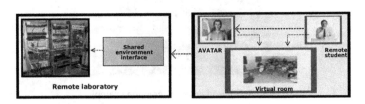

Fig. 7. Immersive lesson example

4   Research Challenges Beyond VICOM

According to the recent "ISTAG SCENARIOS FOR AMBIENT INTELLIGENCE 2010" [7] the evolutionary scenarios will be rooted within three dominant trends[8]:

- pervasive diffusion of intelligence in the space around us, through the development of network technologies and intelligent sensors;
- increasingly relevant role of mobility, through the development of mobile communications, moving from the Universal Mobile Telecommunications System (UMTS) "Beyond 3rd Generation" (B3G);
- increase of the range, accessibility and comprehensiveness of communications, through the development of multi-channel multimedia technologies.

The convergence of biosensors, 4G mobile communication and multi channel multimedia technologies manifests itself as the next frontier of ICT (Information and Communication Technology). An important role will be played by intelligent environments in which complex multimedia contents integrate and enrich the real space.

Within this process two trends are expected to shape the future challenges beyond VICOM: Ambient Intelligence (AmI) and Immersive Virtual Telepresence (IVT).

AmI is an emerging interface paradigm in which the computer intelligence is embedded in a digital environment that is aware of the presence of the users and is sensitive, adaptive, and responsive to their needs, habits, gestures and emotions.

IVT is a new hybrid platform including shared virtual reality environments, wireless multimedia facilities - real-time video and audio – and advanced input devices – tracking sensors, biosensors, brain-computer interfaces. For its features IVT can be considered an innovative communication interface based on interactive 3D visualization, able to collect and integrate different inputs and data sets in a single real-like experience.

A typical first generation IVT system is virtual reality [9](Fig. 8). In VR, using visual and auditory output devices, the user can experience the environment as if it were a part of the world. Further, because input devices sense the operator's reactions and motions, the operator can modify the synthetic environment, creating the illusion of interacting with and thus being immersed within the environment.

IVT, however, is not only a hardware system. According to different authors the essence of IVT is the inclusive relationship between the participant and the synthetic environment, where direct experience of the immersive environment constitutes communication [10]. In this sense, IVT can be considered as the leading edge of a general evolution of present communication interfaces like television, computer and telephone. Main characteristic of this evolution is the full immersion of the human sensorimotor channels into a vivid and global communication experience: IVT provides a new methodology for interacting with information.

For this reason, next generation IVT systems will have an improved focus on the communication capabilities. A possible future IVT application is Mobile Mixed Reality (MMR) [11]. This application foresees the enhancement of information of a mobile user about a real scene through the embedding of one or more information objects within his/her sensorial field. These objects may be part of a wider virtual space – the AmI Space - whose contents can be accessed in different ways and using different media (cellular phones, tablet PCs, PDAs, Internet, etc.).

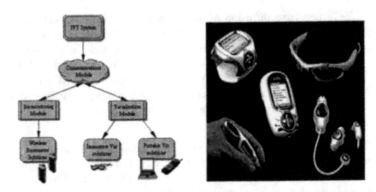

Fig. 8. The IVT system functional architecture and a prototype under development by Motorola

The possibilities offered by MMR are huge. By integrating within a common interface a wireless network connection, wearable computer and head mounted display, MMR virtually enhances users' experience by providing information for any object surrounding them. They can manipulate and examine real objects and simultaneously receive additional information about them or the task at hand.

Moreover, using Augmented or Mixed Reality technologies, the information is presented three-dimensionally and is integrated into the real world. Recently, Christopoulos [12] identified the following applications of MMR:

- Smart signs added to the real world: Smart signs overlaid on user real world may provide information assistance and advertisement based on user preferences.
- Information assistant (or "virtual guide"): The virtual guide knows where the user is, his/her heading, as well as the properties of the surrounding environment; interaction can be through voice or gestures, and the virtual guide can be an animated guide and provides assistance in different scenarios based on location and context information.
- Augmented Reality or Virtual Reality combined with conversational multimedia (or "virtual immersive cooperative environments"): Conversational multimedia can be also added to a VR or an augmented reality scenario, where a user can see the avatar of another user coming into the scene and a 3D video conference is carried on. If we use VR, given the position and orientation information of the first user in the world, the second user can put the first one (or his/her avatar) in a 3D synthetic world.

In general, the IVT perspective is reached through:

- the induction of a sense of "presence" or "telepresence" through multi-modal human/machine communication in the dimensions of sound, vision, touch-and-feel (haptics);
- the widening of the input channel through the use of biosensors (brain-computer interface, psycho-physiological measurements, etc.) and advanced tracking systems (wide body tracking, gaze analysis, etc.).

Typically, the sense of presence is achieved through multisensorial stimula such that actual reality is either hidden or substituted via a synthetic scenario, i.e. made virtual through audio and 3-D video analysis and modelling procedures. In high end IVT systems, multimedia data-streams, such as live stereo-video and audio, are transmitted and integrated into the virtual space of another participant at a remote system, allowing geographically separated groups to meet in a common virtual space, while maintaining eye-contact, gaze awareness and body language. Presence with other people who may be at distant sites is achieved through avatar representations with data about body movement streamed over a high-speed network. Following these premises, a general system functional architecture for a high-end IVT systems should includes three main modules:

- The Visualization Module will use virtual environments and augmented reality to provide totally new users services and interfaces. The research will focus on the characteristics and components of wearable personal virtual reality systems with augmented reality display systems, tracking systems, wireless communications and wearable computing. Wireless communication is needed between components of the system and also between personal augmented reality system and networks services, such as world models and other users or avatars.
- The Biomonitoring Module will give the access to a wide range of biometrics data to support highly individual services. Biosensors are a neural interface technology that detect nerve and muscle activity. Currently, biosensors exist that measure physiological activity, muscle electrical activity, brain electrical activity, and eye movement.
- The Core Module within the system manages the information flows both internally within the software and externally within the environment to allow remote access and interrogation. This model requires unique messaging services that make the IVT database accessible to external authenticated users.

Transforming this vision in reality is not an easy task: the most a technology is complex and costly, the less the user is prone to accept it. Significant efforts are still required to move AmI and IVT into commercial success and therefore routine use. Possible future scenarios will involve multi-disciplinary teams of engineers, computer programmers, and users working in concert. Information on advances in IVT and AmI technology must be made available to the research community in a format that is easy-to-understand and invites participation. Future potential applications of these tools are really only limited by the imaginations of talented individuals.

5 Conclusions

The main target of the VICOM project is the development of two large test-beds in which mobile communication and virtual environment will be merged to offer to the user an immersive experience in a mobile environment. This is only the first step in the achievement of new communication modes pervasive and immersive. The knowledge developed through the VICOM project is carrying out new skills and competences to invest in innovative proposal to the development of full AmI and MMR paradigms.

References

1. VICOM Website: http://www.vicom-project.it
2. F.Vatalaro, "Dalle Telecomunicazioni alla Telepresenza Immersiva", Notiziario Tecnico Telecom Italia, Anno 11, n.3, Dic. 2002, pp. 111-116 (in Italian)
3. G. Riva, P. Loreti, M. Lunghi, F. Vatalaro, F. Davide, "Presence 2010: The emergence of Ambient Intelligence", from "Being There: Concepts, effects and measurement of user presence in synthetic environments", IOS Press 2003, section 4, pp. 59-82, available at: http://www.vepsy.com/communication/volume4/4Riva.pdf
4. Amy L. Murphy, Gian Pietro Picco, and Gruia-Catalin Roman. Lime: A Middleware for Physical and Logical Mobility. In Proceedings of the 21 st International Conference on Distributed Computing Systems (ICDCS-21), May 2001
5. G.P. Picco, S. Cicero, G. Cortese, D. Frey, A. L. Murphy, E. Trevisani, A. Vitaletti - "Software Architecture and Middleware for VICOM: Concepts, API, and guidelines for Application Development"- VICOM project internal document
6. F. Davide, A. Detti, E. Gregori, F. Vatalaro, "Wireless Networking for Virtual Immersive COMmunications: the VICOM Project" in Proc. Personal Wireless Communications (PWC), Venezia, September 2003
7. K. Ducatel, M. Bogdanowicz, F. Scapolo, J. Leijten, and J. C. Burgelma, "Scenarios for ambient intelligence in 2010 (ISTAG 2001 Final Report)," IPTS, Seville 2000
8. F. Davide, P. Loreti, M. Lunghi, G. Riva, and F. Vatalaro, "Communications through virtual technologies," in Advances Lectures on Networking, E. Gregori, G. Anastasi, and S. Basagni, Eds. Berlin: Springer-Verlag, 2002, pp. 124-154
9. G. Riva and B. K. Wiederhold, "Introduction to the special issue on virtual reality environments in behavioral sciences," IEEE Transactions on Information Technology in Biomedicine, vol. 6, pp. 193-7, 2002
10. G. Riva and F. Davide, "Communications through Virtual Technologies: Identity, Community and Technology in the Communication Age," in Emerging Communication: Studies on New Technologies and Practices in Communication, G. Riva and F. Davide, Eds. Amsterdam: Ios Press. Online: http://www.emergingcommunication.com/volume1.html, 2001
11. L. Rosenblum, "Virtual and Augmented Reality 2020," IEEE Computer Graphics and Applications, vol. 20, pp. 38-39, 2000
12. C. Christopoulos, "Mobile Augmented Reality (MAR) and Virtual Reality," Wireless World Research Forum, Stockolm September 17-18 2001

A Framework for Self-organized Network Composition

Cornelia Kappler[1], Paulo Mendes[2], Christian Prehofer[2],
Petteri Pöyhönen[3], and Di Zhou[4]

[1] Siemens AG Communications, 13623 Berlin, Germany
cornelia.kappler@siemens.com
[2] DoCoMo Euro-Labs, Munich, Germany
lastname@docomolab-euro.com
[3] Nokia Research Center, Nokia, Helsinki, Finland
petteri.poyhonen@nokia.com
[4] Siemens AG Austria, A-1210 Vienna, Austria
di.zhou@siemens.com

Abstract. This paper discusses a framework for a flexible, self-organized control plane for future mobile and ubiquitous networks. The current diversification of control planes requires a manual configuration of network interworking. The problem will increase in the future, with more dynamic topologies and integration of heterogeneous networks in a ubiquitous, reactive environment. In this paper we introduce the concept of network composition, a basic, scalable and dynamic network operation to achieve autonomic control plane interworking between Ambient Networks – our approach for next generation networks. We show the architectural components of a generic control plane and its flexible interfaces. With an example on seamless mobility we illustrate how composition can simplify and improve the interworking of future networks.

Keywords: Designing evolvable NGNs, Self-organization for NGN reconfigurability.

1 Introduction

This paper discusses a framework for a flexible, self-organized control plane for future mobile and ubiquitous networks. When looking at the control plane of current networks, i.e. mobile cellular networks and the Internet, we have a very diverse situation. Mobile networks, based e.g. on 3GPP (3rd Generation Partnership Project) standards, have a very powerful, but also inflexible and special-purpose control plane. This means that connecting two such mobile networks, via roaming agreements, results in good interworking, but only for pre-arranged, fixed services such as voice calls, SMS (Short Message Service) or basic data services. Roaming agreements moreover need to be established manually.

On the other hand, the Internet in its current form only has a very basic control plane which enables packet routing between different networks. Hence interworking of networks is easier, but mostly provides best effort data transport. Regarding more

M. Smirnov (Ed.): WAC 2004, LNCS 3457, pp. 139–151, 2005.

advanced features, the global Internet consists of many heterogeneous networks interconnected with varying degrees of trust and cooperation: different control environments are established for services like *Virtual Private Network* (VPN), security, integrated mobility management, *Quality of Service* (QoS), *Network Address Translation* (NAT), and multicast. Hence, connectivity between IP networks is provided, but the control planes of those networks are often not compatible. Network interworking therefore also is typically manually configured.

In the future, more dynamic topologies and heterogeneous networks in a ubiquitous, responsive environment are expected. New kinds of mobile networks will appear, such as *Personal Area Networks* (PANs), *Body Area Networks* (BANs), inter-vehicle networks, and sensor networks, all of which will interwork. The control plane interaction of these networks needs to enable e.g. seamless mobility, end-to-end QoS, integrated security and accounting. For instance, mobility handling is different for a mobile phone, a train network or a BAN. Hence it needs to be negotiated which specific protocols to use and in which way. The configuration of control-plane interaction of such networks needs to become autonomic, because it is a very complex process and yet needs to be realized on-the-fly, and moreover transparently to the user. The owners of future ubiquitous networks often are non-experts and hence cannot be burdened with technical details.

Application scenarios for autonomic configuration of control-plane interaction include

- Automatically established roaming agreements between mobile operators,
- Connecting the access network of a train to access networks along the track,
- Creation of vehicular access networks with changing participants,
- Creation of a users PAN,
- Using the PAN of another user to access the Internet.

We address this problem by introducing a new framework for interworking of the next generation of networks based on work currently under way within the Integrated Project "Ambient Networks" supported by the EU [1]. In this framework, a network is viewed as a *composed* set of *Ambient Networks* (ANs) [2]. We argue that the AN concept will not only ensure the maintenance of the openness, reliability and robustness of the Internet, but will also allow an easy usage of communications services in an increasingly complex mesh of different, particularly mobile, networks. To establish control-plane interaction of networks, we introduce the concept of *network composition*.

We use the following two main concepts as the basis for our framework:

- End systems are seen not as nodes, but as (functionality-reduced) ANs. In the future, end-users will not just own terminal devices, but they will own and operate networks of personal devices like PANs and BANs. The notion of a network is now stretching from single devices over small, user-owned networks to globally operated networks. In this way, we can address the enormous variety of networks in a unified way.

- Network composition is used as the basic, essential operation between AN control planes. Composition enables ANs to cooperate on the control plane; it generalizes and streamlines many existing basic concepts like attaching a node to a network, mobility of nodes and networks (viewed as changing the composition structure) as well as typical inter-operator network agreements.

The remainder of this paper is organized as follows. In Section 2, we discuss related work. In Section 3, we present possible application areas. In Section 4, we describe the concept of composition and how it could overcome today's networking limitations. Section 5 presents an example, and Section 6 draws conclusions and lines out next steps.

2   Related Work

The idea of control-plane interworking in a dynamic or self-organized manner has already been discussed in the literature from different perspectives. The work in [3] propagates a kind of meta-control plane, called knowledge plane, for future intelligent management of the Internet. The knowledge plane has a high-level model of what the network is supposed to do, and relies on tools of Artificial Intelligence and Cognitive Systems. In [4] a self-organizing system is proposed that supports spontaneous information exchange and service deployment in ad hoc networks based on interaction patterns between mobile ad hoc nodes. The paper also states the lack of general self-organizing mechanism for dynamic communication environments like mobile ad hoc to support a stable operating environment for applications. [5] introduces the concept of EgoSpaces that are coordination models and middleware for mobile ad hoc networks to provide means for applications to adapt context changes occurring in dynamic environment. Their design goal is to provide a formal abstract approach to context-awareness and middleware managing an extended notion of context. [6] represents an architecture in which services are continuously evaluating system conditions in a self-organized manner to adjust service placement and capabilities.

The authors in [7] argue one of the main functions of future networks will be information delivery, and the underlying technology needs to disappear from the user's perspective. However future network will also be very diverse, and they will be managed by a large number of independent operators. Hence for transparency of the underlying technology control-plane interworking is required. [8] studies the reasons why IP-based QoS is not widely deployed, and concludes some main reasons for this is lack of integrated control and management, simplicity and measurable guarantees. [9] represents a P2P Wireless Network Confederation (P2PWNC) model, in which a set of administrative domains is providing wireless Internet access to each other's users. The authors aim to replace the human administrator of roaming agreements by Domain Agents (DA), thus eliminating administrative overhead.

While all of these research efforts address many critical issues, they do not fully address the emerging needs of future wireless and ubiquitous networks. They are problem statements, or they are focused on specific environment such as mobile ad-hoc networks. However, these coexisting different environments need to cooperate in the future, which is the main goal of our approach. We need to consider self-organized establishment of QoS, management of user and network mobility and other control functions in highly dynamic heterogeneous networks.

3   Application Scenarios

In this section, we discuss two applications scenarios, which illustrate the concept of network agreements in current and future networks. This will show why a new, generic and autonomous solution is needed for future ubiquitous networks.

Limitation of Current Roaming Agreements. Nowadays, a roaming agreement is established between two or more wireless operators outlining the terms and conditions under which the each operator will provide wireless service to the other's subscribers.

Roaming is usually associated with cellular mobile technologies, such as GSM (Global System for Mobile Communication), but it can also be applied to other type of wireless technologies, such as WLAN (Wireless Local Area Network). For instance, in [10], roaming between 802.11 networks and 3GPP networks is described. In its most simple form, the user of the 802.11 network is authenticated based on the SIM card in the 3GPP network. More advanced interworking, which is still to be defined, will also allow seamless handover between the two technologies, i.e. communications are interrupted when a handover is performed. However, with today's roaming agreements, services are not seamless for handover between operators, even if the handover is within the same technology.

The current concept of roaming agreements between operators is quite limited, since agreements are long-lived and commonly manually established for well-defined services between a pre-known set of commercial operators. Next generation networks however will enable a very large number of flexibly defined services in addition to those already known. These services will be offered by large operators as well as private users, in networks of distinct size from a PAN to a backbone network. There is a need to realize agreements concerning these services between networks. Users are "always on" and services can be accessed anywhere. Networks form dynamically, they move, and flexibly react to the users' needs. Such scenarios can only be handled if roaming agreement establishment becomes more dynamic, flexible and self-organized.

Network Agreements for Next Generation Networks. A future business man is using his PAN while traveling on a train that has its own network. The moving train network needs to establish connectivity with different access networks along the train track that can belong to different operators. The business man connects his entire PAN to the train network in a single step, and enrolls in a videoconference. He would like to go through the videoconference keeping the necessary quality level and without having to deal with on-the-fly configurations and agreements.

To allow the business man to move seamlessly, network functionality such as QoS, mobility, security and charging needs to be realized on-the-fly between train and access networks. Note these functionalities are not independent, as a handover may only be performed if adequate security credentials are provided, and deteriorating QoS may trigger handover etc. Such automatic realization of control-plane interaction between functionalities, and between heterogeneous, moving networks is not possible today, except in special-purpose, functionality- restricted cases.

4   A Framework for Network Composition

In this section, we introduce our new framework for network composition. The goal is to provide a flexible and extensible control plane, which can be composed in a self-organized way without manual intervention. We discuss the different kinds of network agreements and give a framework and architecture for managing the agreements. In the following, we first introduce the notion of network composition. Then we discuss different kinds of composition agreements and how to realize them. Also, the interfaces for network composition and architecture are presented. Here, we focus on the internal architecture to enable a flexible, efficient and extendible composition framework, which is not limited to specific services.

4.1   Ambient Network, Ambient Control Space and Functional Areas

An Ambient Network (AN) consists of one or more network nodes and/or devices. It has a common control plane called Ambient Control Space (ACS). Well-defined access to the ACS is provided to other ANs through the Ambient Network Interface (ANI). An AN has one or more identifiers, it can be contacted via the ANI, and it can compose with other ANs. The AN architecture is schematically shown in Figure 1.

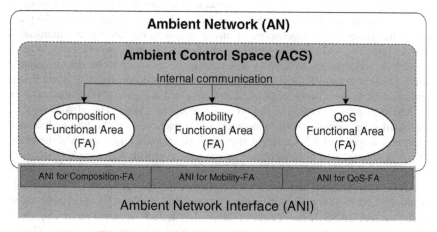

Fig. 1. An example of the Ambient Network Architecture

In the second example of Section 3, the business man's PAN is an AN comprising one or many devices with a joint control of the available resources. The train network is another AN and itself can be composed; e.g. each railway carriage is an AN and all of them are further composed to be a composed train AN with common control and certain edge nodes towards access networks along the track, see also Fig. 2. Each AN in this scenario has its own ACS. The ACS of the PAN is interacting with the ACS of the train AN via the ANI during a discovery and composition phases to gain access to train services and the Internet for all PAN devices.

There are minimal prescriptions how the ACS is realized, or what functionality it supports. It is organized into a number of so-called Functional Areas (FAs), which allow a grouping of topic-related control and management tasks. E.g. there could be a QoS-FA and a mobility-FA. A given FA integrates existing control functions and protocols, e.g. the mobility-FA includes mobile IP and Foreign Agents, however adds means for cooperation between FAs of the same and other ANs for realizing a composition. Particularly, all ANs have a Composition FA that orchestrates the input from all FAs from the same AN for a composition process.

In the second example presented in Section 3, the QoS-FAs of PAN and train network could agree that the train AN takes care of QoS control on behalf of the PAN. Correspondingly the mobility and security-FAs negotiate to transfer mobility and some security control so that the train network is able to do authentication to an access network on behalf of the PAN. As a result of roaming agreement between the train network and the access network, the train network may delegate some specific control functionalities to the PAN, e.g. it may instruct to perform specific priority packet marking for different traffic types to enable the necessary quality level.

4.2   Ambient Network Interface and Generic Ambient Network Signaling

The ANI is an open interface used by ANs to communicate with each other and therefore it is a network-network interface. Its main task is to enable efficient, and consistent message communications among FAs of the ACSs. This communication can take place during the composition negotiation, or inside a composed AN for communication between FAs. The ANI has to integrate existing legacy protocols and interfaces. When a new FA is added to an ACS, ANI will have to be extended to be able to support communication needs. To this end the ANI has a modular structure; each FA is implementing its own portion of the ANI as represented in Figure 1. The instantiation of the ANI may vary according to the ACS, for example a single logical ANI may be distributed over multiple physical network nodes each of them hosting a dedicated instance of a specific FA, or a single physical network node may implement the entire ANI. A distributed ANI implementation can be used to provide for example redundancy or load balancing.

The Generic Ambient Network Signaling (GANS) is the open base set of protocols enabling transport of signaling messages between FAs via the ANI. It is important to emphasize that GANS does not replace standard or de-facto standard protocols, which are used for instance to exchange routing information or for mobility support. GANS is used to exchange information currently not sufficiently covered by generally accepted protocols – e.g. SLA (Service Level Agreement) negotiation, capability exchange, and roaming agreement negotiation. In the example presented in Section 4, the QoS-FAs may start a negotiation using GANS to find out whether they support compatible protocols. When one such protocol has been found and agreed, they may to switch to use that protocol.

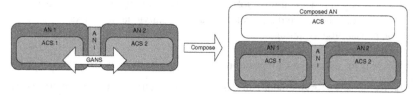

Fig. 2. An example of composition and resulted composed network

Figure 2 represents an example of how two ANs, AN 1 and AN 2, are composing using the GANS protocol to achieve control-plane interaction and correspondingly a composed AN with a joint ACS.

4.3   The Concept of Self-organized Network Composition

The concept of network compositions is introduced to support self-organized control-plane interworking of networks. It enables consistent management over cooperating networks and hides their interconnection details as well as internal structure to the outside. It improves network operation and service efficiency.

The details of control plane interworking between composing networks are fixed in a Composition Agreement. A composition establishment consists of the negotiation and then the realization of a Composition Agreement. Both negotiation of Composition Agreement and its realization should be autonomic i.e. they are usually triggered by internal processes and proceed with minimal user interaction.

Policies play an important role in the composition process. The decision whether to compose is policy-based, the negotiation of the Composition Agreement is policy-based, and the Composition Agreement itself must meaningfully combine the policies of the composing ANs such that the composed AN has its own policies governing future compositions.

4.3.1   Composition Agreements

A Composition Agreements describes all mandatory and optional policies composing ANs agree to follow. A Composition Agreement is created when individual Ambient Networks agree with each other to compose. It is updated when all members of the composed network agree to change it. It exists as long as the composition exists, even when the members of that composition change.

A Composition Agreement is negotiated and created between all FAs of involved ANs. The structure of a Composition Agreement is modular with respect to the FAs. It consists of a general part specifying the basic rules all involved FAs have to follow and a number of subparts referring to agreements between individual FAs. Examples of the content of a Composition Agreement include:

- Identifier of composed AN;
- IP address ranges;
- What resources in which networks are involved;
- Establishing and maintaining QoS of connectivity among individual networks;
- Security associations and trust relations among individual networks;
- Compensation/accounting;
- Common policies to outside and
- The way to realize the Composition Agreement (see more in the next section).

The Composition Agreement can describe a symmetric or an asymmetric sharing of resources, responsibilities, services, duties and permissions between networks involved. An example of a symmetric composition is several BANs composing to set up an ad hoc network, each of them playing a similar role in the composed network. An example of an asymmetric composition is a PAN in a train composing with the train's network to enjoy an entertainment program.

Composition Agreements are expected to often contain off-the-shelf components to improve performance. It is also possible to pre-establish Composition Agreements, or to re-use Composition Agreements negotiated earlier.

4.3.2 Realization of the Composition Agreement

A Composition Agreement can describe more or less tight cooperation of ANs. We loosely distinguish *network integration, control sharing* and *network interworking.*

With network integration, constituent ANs contribute all their logical and physical resources to the composed AN. They give up individual control of some resources and establish a joint ACS. They also hide their own identifiers such that they are not visible individually to the outside. In practice, this means that an AN can only be a member of one such composition at the time. The PAN in our example in Sec. 2 may be one example of network integration, when all its devices (e.g. laptop, PDA, mobile phone) have agreed to give up their individual identities and form a new composed network with a common control plane. Another example is the step-by-step integration and expansion of an infrastructure mobile communication network, where a group of equipment is typically installed and tested as a separate network and then integrated into the existing infrastructure network.

With control sharing, each constituent AN contributes only a part of its logical or physical resources to the composed network but keeps control over the rest. Control of these resources may be delegated to FAs of particular constituent AN, or a joint ACS may be established. An individual network may participate in multiple such compositions in parallel. An example of control sharing are several PANs that build a dynamic ad-hoc network for a meeting, or the delegation of authentication and authorization of the PAN to a train network as represented in our example in Sec. 3.2.

With network interworking, the individual FAs of each constituent AN just coordinate their work. E.g. in roaming agreements, they agree users are always authenticated in the home network.

4.3.3 Composition Functional Area

The Composition Functional Area (C-FA) is an addition to existing control-plane functionality. Its role is the coordination of the FAs of a single AN. It also contains decision logic for running and controlling the composition process. For example, the C-FA collects triggers from other FAs that a composition should be attempted, and takes care all FAs participate in the negotiation and realization of the Composition Agreement. There are minimal assumptions about the ways C-FA may operate, be implemented or managed except its existence. A "Master C-FA" that actively drives the composition process based on policies may be rather straight-forward to implement. Another extreme is a "passive C-FA" that just collects input from other FAs, posts it for others to read, and makes sure it is consistent. However even a passive CA needs to have a logic that drives it to react on certain input. E.g. when it receives a trigger that composition should be attempted, it should make sure a decision is reached on the Composition Agreement in a timely fashion.

4.3.4 Composition Creation

A composition with an active "Master C-FA" could schematically proceed as follows: AN X discovers AN Y, e.g. by receiving a radio beacon, or by user interaction, and learns the identity of AN Y. Upon learning about the discovery, the C-FA of AN X consults a policy data base and finds out composition with AN Y should be attempted. Alternatively, e.g. the mobility-FA could prompt composition by reporting to the C-FA deteriorating quality of the current path. Connectivity for control-plane signaling is established, and usually the security-FA authenticates AN Y by interacting with its peer security-FA in AN Y. Now the C-FA finds out what Composition Agreement it could offer. There may be a pre-established or off-the-shelve Composition Agreement attached to the identifier of AN Y. Otherwise, all FAs need to contribute to the creation of the agreement. The Composition Agreement is offered and negotiated with AN Y. Once the agreement is settled, the security-FA needs to authorize AN Y. Finally, the Composition Agreement is realized.

Three other procedures are needed to realize compositions, namely Composition Extension, which is used by individual ANs to join an existing Composed Network; Composition Agreement Update, which is used by members of a Composed Network to update the Composition Agreement; and Decomposition, which is used by an AN to leave a composition.

5 QoS and Mobility Composition for Self-organized Roaming

This section describes in more detail the usage of the composition framework. In the next generation networks scenario described in Section 3, the business man needs to obtain Internet access for all devices of his PAN in a single step. Moreover, for the video conference he also needs end-to-end QoS, which should be maintained while the train moves, by connecting to different infrastructure access networks along the train track and in stations. The entire process should proceed self-organized with minimal user interaction. Figure 3 illustrates the compositions relevant in this scenario.

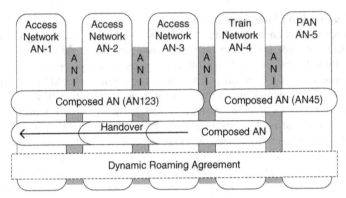

Fig. 3. Compositions in the "business man on a train scenario"

The PAN (AN-5) composes with the train network (AN-4). The composition is of the "control-sharing type". The Composition Agreement states that the train network will provide Internet access to all the devices included in the PAN, independent of its location inside the train, and the train network performs QoS and mobility control as well as some security control on behalf on the PAN, allowing it to be connected without having any perception of movement of the train and its own movement inside the train. By delegating some QoS control, the PAN authorizes the train network to negotiate QoS with the different access networks along the track, and to adapt multimedia sessions to network quality oscillations on its behalf. For mobility control, the train network may e.g. provide a translation service between the care-of addresses of the PAN seen outside the train, and a fixed address that is used by the PAN inside the train. By delegating mobility control, the PAN also authorizes the train network to authenticate it with each access network.

Because the devices of the PAN are composed to an integrated network, the train network only needs to negotiate with one entity, the PAN, rather than with all its constituent devices. While this control plane abstraction is more flexible, it also reduces the signaling load within the train network.

Along the train track, access networks (AN-1, AN-2 and AN-3) compose to provide seamless communication services to the users of the train network, by creating a virtual access network. This composition is of the control-sharing type, in which access networks share logical and physical resources in order to jointly manage QoS and mobility within the virtual access network. The virtual access network delegates access control to the train network, by trusting all users the train network trusts. Joint mobility management may look as follows:

– Access networks may agree to implement a type of inter-network Fast Mobile IP [11], by allowing each network to have pre-configured address information, which reduces control signaling during handover, and eliminates address resolution time.

- They may agree to implement a common Hierarchical Mobile IP [12] scheme, in which a hierarchic of mobility anchor points is jointly used in all access networks.
- Access networks can exchange information about their network capacity and number of clients, or even may combine their context handling schemes in order to allow a wiser decision about the next attachment point.

Regarding QoS, the access networks may agree to establish a consistent QoS management for the composed network. For instance, class-based networks may agree to use the same type of classes, and to exchange information about the usage of resources within each class. This will, for instance, allow admission control to be done only at the edges of the virtual access network and reduce the time required to set up requested QoS levels, contributing to the seamless movement of multimedia sessions.

Since the train may move fast, a network inter-working type composition may be established between the train and the virtual access network. The Composition Agreement describes how QoS and mobility control is handled between the train network and the virtual access network, without having any sharing of control between them. The Composition Agreement presumably is pre-established, since trains of this company frequently travel on these tracks.

In terms of interaction of FAs this composition process could e.g. proceed as follows: Composition may e.g. be triggered by the mobility-FA detecting the virtual access network is in reach. A policy tells it composition should be attempted. This information is relayed to the C-FA, which draws off-the-shelve the well-known Composition Agreement and informs all other FAs composition should be performed according to this agreement.

6 Conclusions

We have explained the new framework of Ambient Networks and composition, which aims to support the ubiquitous, heterogeneous mobile networking vision. We argue that our new abstractions, nodes as networks and network composition, give a more coherent and simplified view for future control architectures. We illustrated the need for self-organized dynamically configurable control planes, particularly for network interworking. The concept of composition aims to provide such interworking.

The composition concept proposed in this paper can include e.g. the TurfNet architecture [13]. TurfNet describes an inter-domain communication mechanism that does not require global network addressing or a common network protocol. Hence, it provides an approach to solve the problem of composition of address spaces and inter-domain routing. The different types of network composition considered by TurfNet, namely horizontal and vertical composition, map to our terms network integration and control sharing / network interworking respectively.

The Ambient Networks approach is essential for ubiquitous environments for several reasons.

- The responsibility for network control functions such as QoS and mobility should not be placed on the end system (edge) alone, especially for limited, wireless devices, possibly without user interfaces, in a highly dynamic environment. With composition, control functions can be explicitly assigned and distributed.
- Mobile networks will need a much larger variety of control plane interworking options than possible with static network agreements and fixed protocol solutions.
- Dynamic internetworking is simplified if the procedure is independent of the nature of the entities involved. It shouldn't matter whether a single device, a PAN or the mobile network of a train (itself containing terminals and PANs) is attached to an access network: An Ambient Network can be a single node, a network, or a network of networks. Composition always proceeds according to the same procedure.
- The configuration of control-plane interaction needs to become an autonomic process, because it is very complex and yet needs to be realized on-the-fly, and moreover transparently to the user.

We have shown a control plane framework, which is extendible based on the concepts of functional areas. Furthermore, we have presented design guidelines for a generic signalling protocol for network composition, which coordinates the individual negations of the FAs. The protocol development is still in very early stage. For the composition, we envisage different degrees ranging from loose interworking over control sharing to network integration.

Several issues have not been detailed in this paper for lack of space, e.g. addressing and discovery of ANs. Our current work within the Ambient Networks project aims to detail and validate the framework presented here.

Acknowledgements. This work is supported by the Ambient Networks Project, partially funded by the European Commission under its Sixth Framework Programme. It is provided "as is" and without any expressed or implied warranties, including, without limitation, the implied warranties of fitness for a particular purpose. The views and conclusions contained herein are those of the authors and should not be interpreted as necessarily representing the official policies or endorsements, either expressed or implied, of the Ambient Networks Project or the European Commission.

References

[1] http://www.ambient-networks.org
[2] N. Niebert, A. Schieder, H. Abramowicz, G. Malmgren, J. Sachs, U. Horn, Ch. Prehofer and H. Karl, „Ambient Networks: An Architecture for Communication Networks beyond 3G", IEEE Wireless Communications, April 2004.
[3] D. Clark, C. Partridge, Ch.s Ramming, and J. Wroclawski, "A Knowledge Plane for the Internet", ACM Sigcomm 2003, Karlsruhe, Germany, Aug. 2003.

[4] K. Herrmann, K. Geihs, and G. Mühl, "A Self-Organizing Infrastructure for Mobile Commerce" KiVS 2003.

[5] C. Julien, G.-C. Roman, "Egocentric Context-Aware Programming in Ad Hoc Mobile Environments", Proc. 10th ACM SIGSOFT Symposium on Foundations of Software Engineering, Charleston, SC, USA, 2002.

[6] S. Graupner, A. Andrzejak, V. Kotov, H. Trinks, "Adaptive Control Overlay for Service Management", 1st Workshop on the Design of Self-Managing Systems (AASMS'03), San Francisco, 2003.

[7] R. O'Connor, S. van der Meer, "Present and future organisational models for wireless networks", Proc. 1st Intl. Symposium on Information and Communication Technologies, Dublin, Ireland, 2003.

[8] L. Burgstahler, K. Dolzer, C. Hauser, J. Jähnert, S. Junghans, C. Macián, W. Payer: "The Missing Pieces for QoS Success: What have we learned, why do we care?" ACM SIGCOMM workshop on Revisiting IP QoS, Karlsruhe, Germany, 2003.

[9] Elias C. Efstathiou, George C. Polyzos: "A peer-to-peer approach to wireless LAN roaming", Proc. 1st ACM Int. workshop on Wireless mobile applications and services on WLAN hotspots, San Diego, CA, 2003.

[10] 3GPP TS 23.234 "3GPP system to Wireless Local Area Network (WLAN) interworking; System description", v2.4.0 Jan. 2004.

[11] R. Koodli, "Fast Handovers for Mobile IPv6", IETF, Internet-Draft, March 2003, work in process.

[12] H. Soliman, et al, "Hierarchical Mobile IPv6 Mobility Management", IETF, Internet-Draft, June 2003, work in progress.

[13] S. Schmid, L. Eggert, M. Brunner, J. Quittek, „TurfNet: An Architecture for Dynamically Composable Networks" Proc. 1st IFIP Int. Workshop on Autonomic Comunication, Berlin, Oct. 2004.

Semantic-Based Policy Engineering
for Autonomic Systems

David Lewis[1], Kevin Feeney[1], Kevin Carey[1],
Thanassis Tiropanis[2], and Simon Courtenage[3]

[1] Trinity College Dublin
{Dave.Lewis, Kevin.Feeney, Kevin.Carey}@cs.tcd.ie
[2] Athens Information Technology Centre
ttir@ait.edu.gr
[3] University of Westminster
courtes@westminster.ac.uk

Abstract. This paper presents some important directions in the use of ontology-based semantics in achieving the vision of Autonomic Communications. We examine the requirements of Autonomic Communication with a focus on the demanding needs of ubiquitous computing environments, with an emphasis on the requirements shared with Autonomic Computing. We observe that ontologies provide a strong mechanism for addressing the heterogeneity in user task requirements, managed resources, services and context. We then present two complimentary approaches that exploit ontology-based knowledge in support of autonomic communications: service-oriented models for policy engineering and dynamic semantic queries using content-based networks. The paper concludes with a discussion of the major research challenges such approaches raise.

1 Introduction

Autonomic Communications shares with the vision of Autonomic Computing the desire to develop systems that provide a level of self-configuration, self-optimization, self-healing and self-protection so freeing the human system administrator from having to understand the changing complexities of the distributed IT systems and networks [kephart03]. Autonomic Computing research has largely focused on the management of computing resources, i.e. information server farms, web servers, GRID applications, while Autonomic Communications studies the application of self management to the network management and control domains. The observation in this paper are inspired by research into autonomic ubiquitous computing, which highlights the challenges of extending human governance of systems to the wider user community and of the self-management of system that are highly heterogeneous and ad hoc. While focusing on the particular needs of autonomic communication, the analysis in this paper therefore focusses on the particular challenges of ubiquitous computing.

The need to provide dynamically adaptive management solutions, which can address the increasing complexity and scale of modern heterogeneous networks and

M. Smirnov (Ed.): WAC 2004, LNCS 3457, pp. 152–164, 2005.

adaptive distributed application services, is a fundamental challenge for autonomic system research. Policy Based Management (PBM) has emerged as an attractive approach for flexibly and dynamically controlling systems, services and network behaviour. In particular, over the last few years it has attracted increasing attention from researchers, industry and standards bodies (e.g. IETF, DMTF and TMForum). However, PBM suffer in two important respects. Firstly, as the managed system scales in complexity, it becomes increasingly complicated to determine the impact of policy changes on system behaviour. This problem arises due to the difficulty in linking policy models, which are usually expressed in specific policy languages, to suitable models of both elemental and emergent composite system behaviour. Secondly, current PBM systems are weak in resolving business- and user-level policies into enforceable system-level policies in a generic and automated way. Such interpretation and resolution usually requires expert mediation by a policy author with considerable domain knowledge. This approach becomes unsustainable as responsibility for the management of resources is increasingly delegated and decentralised, reflecting current organisational trends. The problem is further exacerbated as organisations become integrated with other (partner) organizations in e-commerce value chains, virtual organisations, Internet communities or collaborative projects between organisations. This results in significant increases in both the quantity and heterogeneity of the resources that must be managed by the human administrator.

Therefore a more systematic approach to the development and maintenance of policies is required, one which closely integrates the modelling of the managed system and its behaviour with capturing user goals and resolving them to system-executable policies. We call such a systematic approach *policy engineering*. In this paper we propose supporting the policy engineering process by using *ontology-based semantic models* of the managed system to enable automated reasoning about policy resolution and policy interactions. Such reasoning aims to ease the interaction between people involved in policy engineering for system goverance and the autonomic systems over which they have authority and responsibility.

Within the Semantic Web initiative it has been widely observed that ontological reasoning techniques will only provide true benefits once a sufficiently large body of semantically marked up content is available. Similarly in the context of autonomic management, ontology-driven policy engineering will only be of use in the context of autonomic systems where services and networks possess ontological representations. In the following sections we examine two scenarios where ontologies may play an important part in the functionality of services and networks, and can thus also motivate and support the use of ontology-based policy engineering approaches. The first scenario addresses the use of semantic service composition in ubiquitous computing environments (where a variety of service technologies exist, WSDL-based web services, GRID services, Jini, UPnP etc). The second addresses the use of ontology-based semantics in content-based networking for collecting information over heterogeneous systems. In both scenarios we will discuss the role that can therefore be played by ontology-based semantic models of the system in the use of policies for mediating between human stakeholders and the autonomic computing and communications systems involved.

2   Semantic Service Composition in Ubiquitous Computing Environments

In this section we motivate our research into the problems faced in policy engineering by considering the management of ubiquitous computing systems. Such systems require user-driven, autonomous management solutions [weiser]. This domain provides a level of system heterogeneity and of distribution of policy authoring that focuses our investigation directly onto the problems emerging from applying PBM to complex enterprise and service provider networks. Such systems involve physical environments containing a number of devices and accompanying computing resources, which are all potentially networked via a variety of wireless network technologies.

A key characteristic of this domain is that elements can be dynamically combined in different ways to adaptively satisfy user requirements and operational context for a particular task. Users can roam between ubiquitous computing environments, encountering a wide range of computing resources and services, e.g. WLANs, climate controls, printers, as well as introducing resources of their own, e.g. laptops, PDAs, in-car systems and mobile phones. In managing such environments, therefore, very few assumptions can be made about the homogeneity of resources, of system behaviour or of users' rights, responsibilities and goals. Instead these have to be discovered and interpreted at runtime and dynamically resolved into resource management policies that can be enforced on the different elements currently in use.

The development of systematic techniques for policy engineering in such dynamic, heterogeneous application domains is hindered by the current range of different policy languages and variations in their expressive power. The latter is particularly restricted with respect to expressing the objects that make up the managed systems. Such object specifications typically provide incomplete descriptions of the system, for instance concealing the side effects of policy actions which may result in unexpected policy-driven behaviour. Better tool support is required to handle heterogeneity when capturing and processing system behaviour for PBM.

We propose the use of ontologies to handle these issues in support of policy engineering. We adopt a service-oriented approach to describing the behaviour of systems, thus constraining the range of semantic expressions with which we have to deal. Ontology-based semantics are proposed by the Semantic Web initiative [berners-lee] to overcome some of the problems of heterogeneity and runtime discovery of capabilities in the WWW. Research has only recently begun into the use of ontology-based semantics to described services in the form of semantic web services [owls]. Such a Semantic Web based ontology approach will provide benefits to policy engineering in terms of increasingly widespread expertise, tool availability (e.g. editors, validators) and platform support (e.g. inference engines, repositories).

Our approach for defining semantic services for policy engineering extends that taken by the DAML community for OWL-S semantic web service language [owls]. This uses description logic based ontologies of the Web Ontology Language (OWL) [owl] to define the Inputs, Outputs, Preconditions and Effects of a service (often abbreviated to IOPE) as well as the resources used by that service. However we aim to extend the elementary resource model with a richer one suitable for integration with policy engineering. The premise of semantic web services is that the use of separately authored, but inter-changeable ontology models for describing IOPEs and resources allows inference engines, such as AI planners and matchmakers, to automate the discovery, composition, invocation and monitoring of services [mclraith]. The OWL-S specification, therefore, includes mechanisms for specifying the control and data flow constructs needed to define service compositions, reflecting many of the constructs already well established from the workflow and web service composition domains. This provides the highly adaptive model of the managed system needed for the ubiquitous computing domain, where automated service composition may dynamically generate adaptive applications tailored to a user's task requirements and operational context.

Policies, in their simplest form, are event-condition-action rules. They are regarded as being performed on behalf of subjects, i.e. the person or agent wishing to operate on a resource, and acting on a target, i.e. the resource upon which the subject seeks an operation. Policies have been employed in system access control, defining authorisation rules about whether a particular subject is permitted or denied access to a particular target resource [sloman]. Policies are also being increasingly applied to the management of IP network, e.g. for access control and QoS management [stone]. General purpose policy languages address both authorisation policies and obligation policies, the latter being rules about what and when a particular subject is required to do or not do to a particular target [damianou][uszok][kagal]. Policy languages assist administrators with the task of managing large policy rule sets through abstractions such as roles (used for grouping users) and domains (used for grouping subjects, targets and sub-domains). The engineering of policies is assisted through the reuse of policy specification elements, thereby also encouraging consistent understanding of similar policy rules.

Recent research has begun to exploit ontologies for more extensible expression of subjects and targets as well as exploiting existing inference engines to ease policy engineering problems such as policy authoring, de-confliction and distributed enforcement [tonti]. In our scenario, in order to address problems of detecting and resolving unwanted policy-driven behaviour in complex systems while at the same time automating support for resolving policies provided a wider range of users, the ontology-based approach to specifying policies must be combined with the managed service's semantic service models.

Problems in maintaining a set of accurate policies arise when the managed system's complexity grows beyond the ability of individual policy engineers. Typically this prompts the division of policy authoring between teams thereby increasing in the potential for policy conflicts. Some policy languages can detect and, in certain cases, resolve modal conflicts, e.g. authorisation policies simultaneously denying and permitting access by a particular user role to a particular target resource.

However, application specific conflicts are usually only detectable by manual inspection or on a case by case basis at runtime [lupu], and remain a major technical problem in the engineering of industrial scale policy sets. An administrator's knowledge of certain types of known conflicts can be captured as meta-policies which define rule execution precedents or prevent the authoring of potentially conflicting policies. However, where application specific policy conflicts have not been predicted, conflicts must be detected at runtime by exceptions at policy enforcement points. More problematically, concurrent policies executed at different policy execution points may result, possibly through ill-defined side effects, in mis-use or mis-allocation of resources in a way that is difficult to determine. Static detection of such problems is difficult and requires careful control and observation of different policy settings deployed across the system [dunlop].

In the ubiquitous computing scenario, the dynamic policy conflict detection problem is greatly exacerbated by the fact that the system itself changes greatly over time due to the large number of autonomous devices networking together in an ad hoc fashion to provide users with the application functionality they require. Such ad hoc collections may include devices and resources that users may temporarily bring to a particular ubiquitous computing environment, so a static pre-determination of available resources will be impossible to extract. Furthermore, user applications will not be statically bound to specific environments. Applications will be composed from service components in an ad hoc fashion in reaction to the user's current task requirements and other contextual information such as the user's location, social setting, schedule, authorisations, terminal display capabilities and wireless access links. Thus the authoring of application policies will suffer an even greater level of target system fluidity.

Finally, the fact that the application services and resources involved will be sourced from an un-limitable range of vendors will confound agreements on and conformance to standardised management models [osullivan].

Authoring management policies typically requires administrators to interpret natural language business rules and operational policies set down over time by organisational managers. Though it has been an oft stated objective of policy based management to automate the resolution of business goals into system-level policies, general purpose solutions to this problem are still the subject of active research [beigi][bandara]. This is, at least in part, due to the lack of a mechanism for providing the contextual knowledge needed to unambiguously interpret natural language business goals or user policies into a form that can be resolved into system-level resources.

Organisations own resources such as Wireless LAN, Internet access, email servers, file stores, and printers, the use of which administrators must ensure conforms with managerial goals. In addition, administrators (i.e. operational support staff) at wireless access network service providers will need to ensure that the key radio and network resources of ubiquitous computing environments are managed according to business goals implemented as admission control and QoS provision policies [murray]. Organisations are increasingly decentralising their business processes by flattening hierarchies and moving the responsibility for resource management decisions closer to the users of those resources. This results in policy authoring moving beyond just the sphere of specialised administrative staff and being given to individual team leaders and workers across the organisation.

High-level policy languages allow subject and target specifications to be better aligned with concepts more familiar to the average user. However, few system and network management solutions natively support this level of abstraction, and so some bridging mechanisms are required [hull03a]. Where there is a wide heterogeneity in the elements of the managed system, the task to develop suitable bridging mechanisms becomes prohibitively large. This picture is exacerbated when we consider situations where teams with resource management responsibilities span organisational boundaries, involve individual contract workers or temporary visitors, or loose coalitions, such as many Internet-based communities.

Following the ubiquitous computing scenario to its full potential, application components and wireless resources will be adaptively assembled into disposable compositions on a per user, per task basis [chakraborty][masuoka]. The individual user, not wishing to be made aware of the details of each composition, may provide a set of preferences they wish to apply to their activities in general and to certain circumstances in particular, e.g. 'always select the cheapest available wireless link', 'only use secure applications for business tasks', 'always log activities to my home web site'. These preferences are effectively user defined policies that both control the resources they personally own, but which also can constrain the adaptive behaviour exhibited by the ubiquitous computing environments encountered to outcomes with which the user is comfortable.

These preferences need to be expressed in terms with which the user can relate, in particular in terms that relate to the tasks they wish to perform and the effectiveness or quality of service of the adaptive application generated by the ubiquitous computing environment to support this task [hull03b]. These user-defined policies need to be effectively resolved into system level policies and reconciled with the policies set by other users, teams and administrators responsible for the resources involved.

To establish a semantic service model that supports our policy engineering needs while remaining consistent with mainstream Semantic Web development we propose extending the OWL-S specification to improve the expression of the semantics for managed systems. We must therefore focus on specifying management semantics at two levels. One is at the level of the managed system as an application used by the user, which will typically involve composite services consisting of other composite and/or atomic services. The second level relates the resources used by services, which may participate in many concurrent service compositions.

OWL-S currently specifies a simplistic model of the resources that are used by a service and relatively few examples of its application exist. We propose extending the OWL-S resource ontology so that policies can specify, to varying degrees, how a service makes use of resources in various situations. We assert that the component developer is best placed to provide component-specific meta-policies restricting how service user or administrators can later apply policies controlling the deployed service's resource-related behaviour. In this way we can investigate how ontology-based semantics can be used to capture developer knowledge of a service component's use of resources and how this knowledge can be used in the engineering of system-wide policies.

An initial structure for such an integration of service model, resource model and policy vocabulary centred on a deployable component is suggested in [lewis04a]. This represents a service component's managed behaviour as a rule-based automaton. This is in line with several models for ubiquitous computing components that express behaviour as rule sets [fitzpatrick][terada][owen]. Policy rules local to the service component can be set by the service administrator to govern how a particular deployment of the service component makes use of particular resources. Thus, administration of a service-oriented system is enforced by policies local to services components, rather than by policies that relate generally to the underlying resources. This makes the means of management consistent with the service-oriented principle that resources can only be accessed via well–defined services, in this case a policy-based component management service. It also provides the means for a service component developer to design a more flexible component thus offering wider applicability through a common management mechanism.

Overall, the responsibility falls on the component developer to expose, via the extended semantic service specification, all the interactions between the service and the resources it uses, and to make appropriate control of these interactions possible via component-specific meta-policies. As behaviour rule sets can quickly become unmanageable we are currently examining the use of component behaviour ontologies based on finite state machines to expose just a selected subset of behaviour for policy-based management purposes.

A further restriction of the OWL-S semantic service model is that, in common with many web service models, it is concerned only with the behaviour of application software functions, and does not model the behaviour of the networks that link them. From a management point of view, we are equally concerned with the behaviour of application software and of network resources, and therefore will need to include the latter in our semantic service model, thereby exposing their semantics for inclusion in end-to-end management, e.g. of performance, faults etc. Architectural Description Language (ADLs) [medvidovic] model systems as contracts, which are analogous to semantic services in their use of IOPEs, and connectors that enable contracts to communicate, while still possessing their own distinctive properties, e.g. representing the non-functional features of communication networks and middleware.

In ADLs, connectors are modelled as groupings of contracts. We can therefore model semantic connectors building on OWL-S in a way similar to semantic services, and in particular using the same policy based mechanisms for expressing how the implementation of a semantic connector interacts with the communication resources it uses. This will provide us with a single, simple mechanism for modelling a wide range of integrated service and network resources that may be dynamically configured for different applications. The connector abstraction also serves as the integration interface for more complex resource models that are not suited to expression using description logic ontologies.

In our ubiquitous computing application, composite service may consist of services and connectors supporting a number of resources on user terminals, wireless access networks and application services available in a particular environment. The service composition represents the semantics of the managed system to which user level policies need to be applied. By integrating the ability to enforce system level policies into the semantic descriptions of services and connectors, we are able to reason about differences in models within the policy resolution process. We can therefore exploit the semantics implicit in a service composition, linking the composite service and its IOPEs to the IOPEs of the constituent services, and thus, at the level of atomic service, to resource-level policies.

3 Knowledge Based Networking

The Semantic Web initiative has encouraged research into how ontology-based queries can be resolved in a distributed peer-to-peer manner between agents holding information with heterogeneous RDF-based semantics that are distributed over the web [cai][stuckenschmidt][tempich]. In [lewis04b] we outline a Semantic Query Based Network (SBQN) service that extends such distributed querying using Content Based Networks (CBN) to provide a Collaborative Information Service. CBNs use a publish-subscribe message delivery paradigm, but with message routing based on filters applied to message content. This provides more flexibility than routing based on a set of predefined message types [pietzuch][carzaniga].

The architecture for this service aims to securely and flexibly support the acquisition and dissemination of information between members of collaborative groups working across the Internet. It uses persistent ontology-based queries for defining the information being sought and shared, so that the range of supported application domains automatically reflects the ever-expanding range of domain ontologies that will be published for use in the Semantic Web. The SBQN service is also used to support the autonomic management and knowledge management needs of the network itself. The Collaborative Information Service used a SBQN to connect a number of web servers. Each server supports a different set of resource types, where resources are anything with a URI, the meta-data of which is described using OWL. Servers provide standard HTTP pull access to resources, and the SBQN supports push capabilities for resource meta-data. Permissions to advertise and subscribe to meta-data are managed by the user community.

In the autonomic communication domain, we propose that the SBQN could be used for providing fine-grained dynamic collection of network status information. In the first instance such an application could use ontological models of existing network element MIBs [vergara], acting as a flexible management notification delivery service. However, the application of the SBQN becomes more interesting when extended to multiple administrative domains, including the user or customer domain and a dynamically formed chain of network and application service providers. Such situations may become increasingly dynamic as users may more easily select wireless access networks and wired backbone transport is commoditised and providers are switched dynamically. Impose on this an increase in customer applications that operate in the context of virtual organisation, then the need for flexible multi-organisational mechanisms to resolve access control policies on information mediated by the SBQN becomes apparent.

Role-based access control requires a priori agreement on roles and thus its use is limited in such fluid organisations structures. We have therefore established a more flexible abstraction for policy-based access control, where communities, rather than roles, become the central abstraction used to specify access control. In this case communities may represent resource management teams in organisations participating in a virtual organisation as well as the service provider administrators in any service provision chain they use. The power of this approach is that a community may mandate authority to both access resources and to author new access control policies for certain resources to a sub community. In this way the authoring of policies is distributed through a virtual organisation to the group considered best qualified to make those policies. An organisation thereby can organically grow and change, perhaps starting from a single group with all authority, and decomposing into subgroups with mandated authority based on specialisation of skills as the need arises naturally in the organisation's lifecycle.

Conflicts between policies authored by separate groups are automatically reported to the nearest mutual parent of the two groups. This group, having mandated policy authoring authority, is the best placed to decide how to resolve any resulting conflict. The resolution of such conflicts depends on a semantic model consisting of linked directed, acyclical graphs that represent the community structure, the authorisation dependencies between resources and the authorisation dependencies between actions that can be performed on those resources. Specific ontology-based models for these are currently being devised in alignment with the service and resource models outlined in the previous sections.

The semantics of the community-based policy abstraction and of how authority is mandated to sub-communities has been established [feeney]. A community policy management system has been implemented using Ponder, an existing policy based framework [damianou][sloman], and is found to operate satisfactorily. This implementation has been applied to the access control for a CVS code repository being used by an Internet community. Usage tests with that community are currently underway. This will provide us with initial user acceptance results with respect to the use of community-based access control.

4 Discussion and Further Work

These two scenarios presented above provide some motivation for why ontology-based semantics will be important in the modelling of adaptive networked systems, while also demonstrating how such semantics can potentially support more intelligent interaction between the people and autonomic systems. This extends the idea of a knowledge plane for the Internet [clark], to that of a semantic representation for services, for network links (as connectors) and for the possible constraints over adaptivity that can be imposed on the underlying resources. We thus emphasise the point that as autonomic management is essentially human governance resulting in the constraint of adaptive behaviour using policies, we must address the semantics of both adaptive networks and adaptive application software in relating such policies to the expected human experience.

This approach however, leaves many open questions relating to the limits of semantic based reasoning in the context of adaptive, networked systems. Currently standardised ontologies are based on description logic, which is soon to be complemented with semantic rules [horrocks]. Though the latter will obviously aid in the representation of policies using ontologies, it is far from clear that all forms of policy based management can be addressed with these ontological logics. For instance in [kephart04], policies are categorised into action policies, goal policies and utility logic. Though a combination of OWL and SWRL may go some way to being able to reason about the former (and then only with the support of ontologies for temporal logic), other logics suitable for feeding optimization algorithms may be needed. This in turn will require an extensible, modular structure for reasoners that is embedded in the network, similar to existing semantic application toolkits [oberle], but which is itself subject of autonomic management. For example, in the SBQN we envisage nodes dynamically subscribing with queries related to logic problems which they encounter, in order to locate suitable downloadable code to conduct the required reasoning. Equally, ontologies capturing mappings between concepts in separate domain ontologies that appear in user queries can also be sought and obtained by SBQN routing nodes using the SBQN service.

The SBQN architecture raises several issues that require further investigation in order to assess usability and scalability of this architecture for deployment on the Internet. We must perform a more detailed assessment of the performance possible with existing ontology-based matching algorithms, though in the long term we expect that optimised software and hardware support for OWL will emerge driven by its potential popularity, as has already happened for XML processing. One possible optimisation that will reduce the reasoning load on SQBN nodes will be to decompose ICS queries based on known routes prior to submitting them as subscription queries to the SQBN.

In general, further experimentation will be required to evaluate the scalability and performance of such knowledge based networking against variations in the numbers of information sources, sinks, advertisements, subscriptions and client joins/leaves. More challenging is the need to assess scalability against growth in the number and scope of ontology domains, ontology encoded logics and ontology mappings.

As pointed out in [barrett], within any realistic business scenario, policy authoring is a challenging collaborative activity. The community-based policy management approach we present goes some way to addressing the identification and resolution of conflicts between policies developed by different groups within fluid organisations, however the scheme does not yet exploit the full potential of ontology-based policy semantics for dealing with uncertainty about resources and identity as proposed in [kagal].

References

[bandara] Banadara, K., Lupu, E, Moffett, J., Russo, A., "A Goal-Based approach to Policy Refinement", in proc of 5th IEEE International Workshop on Policies and Distributed Systems and Networks, IEEE, 2004, pp 229-239

[barrett] Barrett, R., "People and Policies: Transforming the Human-Computer Partnership", in proc of 5th IEEE International Workshop on Policies and Distributed Systems and Networks, IEEE, 2004, pp 111-116

[beigi] Beigi, M., Calo, S., Verma, D., "Policy Transformation Techniques in Policy-based Systems Management", in proc of 5th IEEE International Workshop on Policies and Distributed Systems and Networks, IEEE, 2004, pp 13-22

[berners-lee] Berners-Lee, T., Hendler, K., Lassila, O. (2001), 'The Semantic Web', Scientific American, pp 35-43, Issue 284 (3), 17th May 2001

[cai] Cai, M., Frank, M., "RDFPeers: A Scaleable Distributed RDF Repository based on a Struuctred Peer-to-Peer Network, in Proc. of World Wide Web Conference 2004, 17-22 May 2004, New York, NY, USA

[carzaniga] Carzaniga et al, "Design and Evaluation of a Wide-Area Event Notification Service", In ACM Transactions on Computer Systems, Vol. 19(3), 2001, pp. 332-383.

[chakraborty] Chakraborty, D., Perich, F., Joshi, A., Finin, T., Yesha, Y., "A Reactive Service Composition Architecture for Pervasive Computing Environments", In 7th Personal Wireless Communications Conference (PWC 2002). Singapore. October. 2002.

[clark] Clark, D., Partridge. C., Ramming, J.C., Wroclawski, J.T. "A Knowledge Plane for the Internet", in Proc. of SIGCOMM'03, 25-29 August 2003, Karlsruhe, Germany

[damianou] Damianou, N., Dulay, N., Lupu, E., Sloman, M., (2001) "The Ponder Policy Specification Language", Proc. Policy 2001: Workshop on Policies for Distributed Systems and Networks, Bristol, UK, 29-31 Jan. 2001, Springer-Verlag LNCS 1995, pp. 17-28

[dunlop] Dunlop, N., Indulska, J., Raymond, K. (2003) "Methods for Conflict Resolution in Policy-based Management Systems", IEEE 7th International Enterprise Distributed Object Computing Conference (EDOC'03), Sept 16 - 19, 2003, Brisbane, Australia, pp 98-111

[feeney] Feeney, K., Lewis, D., Wade, V. "Policy-based Management for Internet Communities", in proc of 5th IEEE International Workshop on Policies and Distributed Systems and Networks, IEEE, 2004, pp 23-34

[fitzpatrick] Fitzpatrick, A., Biegel, G., Clarke, S., Cahill, "Towards a Sentient Object Model" In proceedings of the Engineering Context-aware Object-Oriented Systems and Environments workshop at OOPSLA 2002

[horrocks] Horrocks, I., Patel-Schneider, P., Boley, H., Tabet, S., Grosof, B., Dean, M. (2003), "SWRL: A Semantic Web Rule Language Combining OWL and RuleML", version 0.5, 19th November 2003, http://www.daml.org/2003/11/swrl/

[hull03a] Hull, R., Kumar, B., Lieuwen, D. (2003), "Towards federated Policy Management", IEEE 4th International Workshop on Policies for Distributed Systems and Networks, June 04 - 06, 2003, Lake Como, Italy, pp 183-196

[hull03b] Hull, R., Kumar, B., Lieuwen, D., Patel-Schneider , P., Sahuguet , A., Varadarajan, S., Vyas, A. (2003), "Policy-based System for Personalized and Privacy-conscious User Data Sharing", Bell Labs Technical Memorandum, February, 2003.

[kagal] Kagal, L., Finin, T., Joshi, A., "A Policy Language for A Pervasive Computing Environment", IEEE 4th International Workshop on Policies for Distributed Systems and Networks, June 04, 2003

[kephart03] Kephart, J., Chess, D., "The Vision of Autonomic Computing", IEEE Computer, Jan 2003, pp 41-50.

[kephart04] Kephart, J., Walsh, W., "An Artificial Intelligence Perspective on Autonomic Computing Policies", in proc of 5th IEEE International Workshop on Policies and Distributed Systems and Networks, IEEE, 2004, pp 3-12

[lewis04a] Lewis, D., Conlan, O., O'Sullivan, D., Wade, V. "Managing Adaptive Pervasive Computing using Knowledge-based Service Integration and Rule-based Behavior", in Proc. of 2004 IEEE/IFIP Network Operations and Management Systems, pp 901-902 [lewis04b] Lewis, D., Feeney, K., Tiropanis, T., Courtenage, S., "An Active, Ontology-driven Network Service for Internet Collaboration", in Proc of Semantic Web for Web Communities workshop, 2004, Valencia, Spain.

[lupu] Lupu, E.C, Sloman, M. (1999), "Conflicts in Policy-Based Distributed Systems Management", IEEE Transactions on software engineering, vol. 25, no. 6, November 1999. pp 852-69.

[masuoka] Masuoka, R., Labrou, Y., Parsia, B., Sirin, E., "Ontology-Enables Pervasive Computing Applications", IEEE Intelligent Systems, Sept/Oct 2003, pp 68-72

[mclraith] McIlraith, S.A., Son, T.C., Honglei Zeng, H. (2001), 'Semantic Web Services', IEEE Intelligent Systems, 16(2), March/April 2001

[medvidovic] Medvidovic, N. Taylor, R.N., (1997) "A framework for classifying and comparing architecture description languages", Proc. 6th. European Software Engineering Conference/Proc. 5th. Symposium on the Foundations of Software Engineering, Zurich, Switzerland, Sept. 1997, pp 60-76.

[murray] Murray, K., Mathur, R., Pesch, D., "Adaptive Policy-Based Access Management in Heterogeneous Wireless Networks", in Proc of IEEE WPMC, Vol. 1, pp 325-329, Oct 2003

[oberle] Oberle, D., Staab, S., Volz, R., "An Application Server for the Semantic Web", in Porc of WWW'04, May 17-22, 2004, New York, USA, pp. 220-221.

[osullivan] O'Sullivan, D., Lewis, D., "Semantically Driven Service Interoperability for Pervasive Computing", Proceedings of the 3rd ACM International Workshop on Data Engineering for Wireless and Mobile Access, San Diego, CA, USA, 19th September 2003, pp 17-24

[owen] Owen, T., Rathke, J., Wakeman, I., Watson, D, "Implementing Policies in Programs using Labelled Transition Systems Tim Owen Julian", In Cosener's House Multi-Service Networks Conference, 2002, June 29, 2002

[owl] W3C (2003) Ontology Web Language, http://www.w3.org/2001/sw/, Visited Apr 2003

[owls] "OWL-S: Semantic Markup for Web Services", The DAML Service Coalition, http://www.daml.org/services/, October 2002

[pietzuch] Peter R. Pietzuch and Jean Bacon. Peer-to-Peer Overlay Broker Networks in an Event-Based Middleware. In H. Arno Jacobsen, editor, Proceedings of the 2nd International Workshop on Distributed Event-Based Systems (DEBS'03), ACM SIGMOD, San Diego, USA, June 2003. ACM.

[sloman] Sloman, M., Lupu, E., "Security and Management Policy Specification", IEEE Network, vol.16 No. 2, March/April 2002. pp 10-19.

[stone] Stone, G.N., Lundy, B., Xie, G.G. (2001), "Network Policy Languages: A survey and a new approach", Network, IEEE , Volume: 15 , Issue: 1 , Jan.-Feb. 2001, pp10 - 21

[stuckenschmidt] Stuckenschmidt, H., Vdovjak, R., Houben, G.J., Broekstra, J., "Index Structures and Algorithms for Querying Distributed RDF Repositories", in Proc. of World Wide Web Conference 2004, 17-22 May 2004, New York, NY, USA

[tempich] Tempich, C., Staab, S., Wranik, A., "REMINDIN': Semantic Query Routing in Peer-to-Peer Networks Based on Social Metaphors", in Proc. of World Wide Web Conference 2004, 17-22 May 2004, New York, NY, USA

[terada] Terada, Tsutoma, Tsukamoto, M., Hayakawa, K., Yoshihisa, T., Kishino, Y., Kashitani, A., Nishio, S., "Ubiquitous Chip: A Rule-Based I/O Control Device for Ubiquitous Computing", in Proc Pervasive, 2004, LNCS 3001, pp. 238-253, 2004

[tonti] Tonti, G., Bradshaw, J.M., Jeffers, R., Montanari, R., Suril, N., Uszok, A., "Semantic Web Languages for Policy Representation and Reasoning: A Comparison of KAoS, Rei, and Ponder" Proceedings of 2nd International Semantic Web Conference (ISWC2003), October 20-23, 2003, Sanibel Island, Florida, USA

[weiser] Weiser, M. (1991), "The Computer of the 21st Century", Scientific American, vol. 265, no.3, September 1991, pp 66-75.

[uszok] Uszok, A., Bradshaw, J., Jeffers, R., Suri, N., Hayes, P., Breedy, M., Bunch, L., Johnson, M., Kulkari, S., Lott, J. (2003), "KAoS Policy and Domain Services: Toward a Description-Logic Approach to Policy Representation, Deconflictions, and Enforcement", IEEE 4th International Workshop on Policies for Distributed Systems and Networks, June 04 - 06, 2003, Lake Como, Italy, pp 93-98

[vergara] de Vergara, J.E.L., Villagra, V.A., Berrocal, J., "Applying the Web Ontology Language to Management Information Definitions", IEEE Communications Magazine, special issue on XML Management, Vol. 42, Issue 7, July 2004, pp. 68-74. ISSN 0163-6804

Dynamic Self-management of Autonomic Systems: The Reputation, Quality and Credibility (RQC) Scheme⋆

Anurag Garg, Roberto Battiti, and Gianni Costanzi

Dipartimento di Informatica e Telecomunicazioni,
Università di Trento,
Via Sommarive 14, 38050 Povo (TN), Italy
{garo, battiti}@dit.unitn.it
gianni.costanzi@studenti.unitn.it

Abstract. In this paper, we present a feedback-based system for managing trust and detecting malicious behavior in autonomically behaving networks. Like other distributed trust management systems, nodes rate the interactions they have with other nodes and this information is stored in a distributed fashion.

Two crucial insights motivate our work. We recognize as separate entities the trust placed in a node, *reputation*, and the trust placed in the recommendations made by a node, *credibility*. We also introduce the concept of quality of a trust rating. Together, these two factors enhance the ability of each node to decide how much confidence it can place in a rating provided to it by a third party.

We implement our scheme on a structured P2P network, Pastry, though our results can be extended to generic autonomic communication systems. Experimental results considering different models for malicious behavior indicate the contexts in which the RQC scheme performs better than existing schemes.

Keywords: Trust management, reputation, quality, credibility, autonomic systems, peer-to-peer systems.

1 Introduction

Autonomic systems aim at incorporating methods to monitor their dynamic behavior and to react in automated ways, leading to self-awareness, self-management, self-healing and self-improvement. This work explores how ideas of automated reputation schemes that have originated in other contexts (e.g. e-commerce) can be modified and applied for the soft enforcement of rules of behavior in specific autonomic communication systems based on distributed control. In particular, this work proposes a novel scheme to self-manage trust in autonomic systems using Peer-to-Peer (P2P) environments as an example setting.

Many autonomic systems implicitly place a certain trust in participants, assuming that they will follow certain guidelines for "fair-use". But due to the distributed nature of such systems, breaking these guidelines can go undetected by the system as a whole

---

⋆ This work was supported by the Provincia Autonoma di Trento through the WILMA Project.

M. Smirnov (Ed.): WAC 2004, LNCS 3457, pp. 165–178, 2005.

and does not result in any significant penalties to misbehaving components. This can result in poorer quality of service and, in the worst case, outright service denial.

For any system that is open and anonymous, imposing barriers for entry is not an acceptable solution. A reputation-based trust management system offers a better solution. These systems rely on the dissemination, throughout the network, of trust information gathered through transactions between nodes. In this way nodes can build knowledge about nodes with whom they have never interacted before and use this information to decide whether to interact with new nodes. But relying on information from third parties also makes the system vulnerable to manipulation through false complaints or false praise.

We present a reputation-based system that is robust against false ratings and at the same time helps "good" nodes to avoid interacting with "malicious" nodes. Along with node Reputation (R), we use Quality (Q) and Credibility (C) to provide a richer and more robust trust management system called the **RQC** system. In this way, the paper contributes to the understanding of how trust information should be aggregated and how much credence should be attached to reported trust values by other nodes in the network.

The rest of this paper is organized as follows. In the next section we review existing work on trust management in P2P systems. In Sec. 3 we present our solution. In Sec. 4 we present our experimental results and we conclude in Sec. 5.

2   Related Work

Initial efforts at trust management in electronic communities were based on centralized trust databases. The eBay rating system used for choosing trading partners where each participant in a transaction can vote $(-1, 0, 1)$ on their counterpart, the Amazon customer review system and the Slashdot self-moderation of posts [1] are all systems where the ratings are provided by nodes but are stored in a central database. Many such reputation systems have been studied in the context of online communities and marketplaces [2, 3, 4].

In true P2P environments, the storage of trust ratings also needs to be done in a distributed fashion. Aberer and Despotovic introduced such a scheme [5] using a decentralized storage system P-Grid to store and retrieve trust information. Peers can file complaints against each other if they feel the node has behaved maliciously. All complaints made by and complaints about a given node, are stored at other nodes called agents. The mechanism is made robust by keeping multiple copies of reports at different agents. When a node wishes to interact with another node, it sends messages querying trustworthiness of the other node to random nodes in the network, which are routed to appropriate agents. Two algorithms are described to compute the trustworthiness. The first relies on a simple majority of the reporting agents' decisions and the second checks the trustworthiness of the reporting agents themselves and disallows any reports coming from untrustworthy agents.

Cornelli et. al. [6, 7] propose a mechanism built on the Gnutella network, where a node uses a secure polling mechanism to ask its neighbors about interactions they may have had with a specific node to gauge its trustworthiness. The scope of the messages querying trust is limited by the Gnutella architecture design. Their work is directed at file-sharing networks and the objective is to find the most trusted node that possesses a given resource and they focus on vote aggregation on incorporating voter credibility and on ensuring the integrity of trust reports as they pass over the insecure network.

Kamvar et. al. [8] use a different approach and assume that trust is transitive. There-fore, a node weighs the trust ratings it receives from other nodes by the trust it places in the reporting nodes themselves. Global trust values are then computed in a distributed fashion by using a trust matrix at each node. Successive iterations involving exchange of trust values with neighbors and re-computation of the matrix. Trust values asymp-totically approach the eigenvalue of the trust matrix, conditional on the presence of pre-trusted peers that are always trusted.

Buchegger et. al. [9] propose a modified Bayesian approach to trust. Like Dami-ani et. al. they separate a node's reputation (performance in the base system such as file-sharing, routing, distributed computing etc.) and its credibility[1] (performance in the reputation system). In their solution, only information on first-hand experiences is published by nodes. This information is used by other nodes to construct their own reputation and credibility data structures for other nodes. Reputation data is also aged giving less weight to evidence received in the past.

3 The RQC (Reputation, Quality, and Credibility) Approach

In this section, we describe the RQC algorithm. In RQC, each node is assigned M score managers and all transaction involving the node are reported to each of its score managers[2]. The score managers aggregate trust information for the node to construct its global trust value or *reputation*. They also respond to rating requests by nodes wishing to transact with that node. A quality value is attached to all trust ratings by the sender (i.e., a node reporting on a transaction or a score manager responding to a trust query). The recipient of these ratings (the score manager and the querying node respectively) weighs them using this attached quality value and the credibility of the sender. Sender credibility is calculating by comparing its reported trust rating with the average value computed at the recipient. A sender that reports values that diverge from the calculated average sees its credibility go down whereas a sender whose reported values agree with the average sees its credibility go up. In the following sections we explain this process in greater detail.

3.1   Local Opinion and Opinion Quality

Each node i maintains an average opinion O_{ij}^{avg} of the behavior of all nodes j with which it has had an interaction. O_{ij}^{avg} can be interpreted as node i's estimate of the probability

---

[1] They call it trust.

[2] Since the score managers are also nodes in the network, we use the term "node" to refer exclusively to a node when it is not acting in its capacity as a score manager.

that node j will behave honestly during a transaction. After each interaction with j, i updates O_{ij}^{avg} and sends the updated value to all the score managers responsible for j.

Along with O_{ij}^{avg}, the node i also stores the number of interactions, N_{ij}, it has had with j and the variance s_{ij}^2 in the behavior of j. Thus, the average local opinion is computed as follows:

$$O_{ij}^{avg} = \frac{\sum_k O_{ij}^k}{N_{ij}} \tag{1}$$

where O_{ij}^k is node i's opinion of its kth interaction with j. Each O_{ij}^k takes a value in $[0, 1]$ and represents node i's satisfaction with node j's behavior during the transaction.

When node i sends its updated average opinion about j (O_{ij}^{avg}) to the score managers, it also sends an associated quality value, Q_{ij}. Q_{ij} represents the quality node i attaches to the opinion information it is sending to the score managers and lies within $[0, 1]$. Q_{ij} enables a node to express the strength of its opinion. Its value could depend on the context of the interaction as well as on past transaction history. In our current implementation, quality values are computed solely on the basis of the number of interactions and the variance in the opinion.

We assume that j's trust behavior is a normally distributed random variable. Through interactions with j, node i makes observations of this random variable resulting in a sample. The sample mean and standard deviation are then simply O_{ij}^{avg} and s_{ij}.

The quality value of the opinion (Q_{ij}) is defined as the confidence level that the actual mean trust rating for a node lies within the confidence interval:

$$O_{ij}^{avg} \cdot (1 \pm \frac{r}{100}) \tag{2}$$

where r is a system parameter that denotes the size of the confidence interval as a percentage of the sample mean. We experimented with various values of r, ranging from 5 to 30 and found that a confidence interval of 10% of the sample mean (and thus $r = 10$) resulted in the best performance. Using too high a value for r produced useless quality values, as large variations in node behavior were allowed without any decrease in the quality. Similarly, too low a value of r resulted in excessively low quality values.

Since the actual mean and standard deviation are unknown, we used the *Student's t-distribution* to compute the confidence levels. Note that the usual idiom is inverted here in that we know the interval and wish to compute the probability that the actual mean lies within the interval as opposed to normal practice where confidence level is known and the required interval is computed.

The t-value for the *Student's t-distribution* is given by the following equation:

$$t = \frac{r}{100} \cdot \frac{O_{ij}^{avg} \cdot \sqrt{N_{ij}}}{s_{ij}} \tag{3}$$

And the quality value is computed as:

$$Q_{ij} = 1 - B\left(\frac{(N_{ij} - 1)}{(N_{ij} - 1) + t^2}; \frac{1}{2} \cdot (N_{ij} - 1), \frac{1}{2}\right) \tag{4}$$

where B is the *Incomplete Beta Function* defined as $B(z; a, b) \equiv \int_0^z u^{a-1} \cdot (1 - u)^{b-1} \, du$.

Thus an opinion is of greater quality when the number of observations on which it is based is larger and when the interactions have been consistent (resulting in a smaller variance). When the number of observations is high but they do not agree with each other, the quality value is lower.

When a node has had only one interaction with the other node, equations 3 and 4 cannot be used since sample variance is undefined for a sample size of one. Instead, a default quality value of 1 is used in this case. If a lower value, such as 0.5, is used, the opinions sent by malicious nodes during the initial interactions (when there is little credibility information[3]) would overwhelm the opinions of the good nodes, as malicious nodes would always report a quality value of 1 for their opinions.

After each interaction, both participating nodes report their updated opinions along with the associated quality values to the score managers responsible for their counterpart. The inclusion of quality in the message sent to the score manager allows the score manager to gauge the how much confidence the node itself places in the rating it has sent.

3.2 Computation of Reputation at the Score Manager

The underlying DHT structure designates M score managers for each node in the network. Since the score managers are selected from nodes within the network itself, each score manager is responsible for M nodes on average. This provides the reputation system with redundancy so that the failure of a few score managers does not affect the trust management system.

A score manager receives an averaged opinion whenever a node it is responsible for is involved in a transaction. This opinion is sent by the other node involved in the transaction. The individual opinions from successive transactions are aggregated to form the *reputation* of a node. The score managers therefore represent the global, system-wide view of a given node's behavior, updated every time a new opinion is received.

Along with the aggregate reputation, R_{mj} for a node j, a score manager, m, also stores the number of opinions it has received about j, N_{mj} and the variance s^2_{mj} in the reported opinions. This information is used to compute the quality value that a score manager attaches to a reputation value. The computation of the quality of reputation values at a score manager is similar to that of the individual nodes' computation of quality values for their averaged opinions described in equations 3 and 4. So, the quality of a reputation value is simply the confidence level that the actual mean reputation of the node is within $r\%$ of the sample mean.

If the reputation value of a node at the score manager has been calculated using the opinion of a single voter only, a quality value of 1 is returned. The reason for this is the same as described in the previous section.

3.3 Retrieval of Trust Information

When a node wants to know the reputation of another node before interacting with it, it locates the M score managers for the node by using the DHT substrate and asks them for the reputation of the node in question. Each score manager responds with a

---

[3] The role of credibility is explained in Section 3.4.

reputation value and an associated quality value. The node then computes the average reputation for the node in question, R_{ij}^{avg} using the quality values and the credibility values of the score managers since a score manager itself may also be malicious and send the wrong reputation values. In this way, multiple score managers allow the system to cope with malicious behavior.

3.4   Credibility

C_{ij} or the credibility of node j in the eyes of node i is the confidence node i has in node j's opinions about other nodes and contributes to the weight node i gives to the opinions expressed by node j and to the reputation values furnished by node j in its capacity as a score manager. A single credibility value is used for a node for both of its roles as a reporting node and a score manager.

Every node stores the credibility rating for each node (or score manager) that has sent it an opinion value (or a reputation value). The credibility rating is updated every time a node reports an opinion or reputation. In addition, when the score manager updates the credibility of a node it uses the quality value furnished by the node to decide the amount of modification in the trust value. This is because a node should not be penalized for an incorrect opinion that was based on a small number of interactions and/or a large variation in experience where this was explicitly stated by the reporting node through a low quality rating. Credibility values are not shared with other nodes and are used simply to weigh the responses received from other nodes and always lie within the range $[0, 1]$.

3.5   Putting It All Together

A score manager uses the quality value sent by a reporting node and the credibility of the reporting node to compute the average reputation of a node. The reputation of a node j is computed at the score manager m as follows:

$$R_{mj} = \frac{\sum_i O_{ij}^{avg} \cdot C_{mi} \cdot Q_{ij}}{\sum_i C_{mi} \cdot Q_{ij}} \qquad (5)$$

where R_{mj} is the aggregated reputation of node j, C_{mi} is the credibility of node i according to the score manager m, O_{ij}^{avg} is the average opinion of j reported by i and Q_{ij} is the associated quality value reported by i. The score manager only keeps the latest value of O_{ij}^{avg} reported by each node i. Thus the score manager gives more weight to ratings that are considered to be of a high quality and that come from nodes who are more credible in the eyes of the score manager.

In the case of reputation retrieval, a node aggregates the responses from the reporting score managers using the reported quality value and the stored credibility value of the reporting score managers. The aggregation is performed in exactly the same way as shown in equation 5 except that the reputation values R_{mj} are aggregated instead of the opinions O_{ij}^{avg}.

$$R_{ij}^{avg} = \frac{\sum_y R_{yj} \cdot C_{iy} \cdot Q_{yj}}{\sum_y C_{iy} \cdot Q_{yj}} \qquad (6)$$

where R_{yj} is a reputation value received from a score manager y about node j.

When a node reports an opinion to a score manager for the first time, its credibility is set to 0.5. Thereafter, every time it reports an opinion on any of the nodes the score manager is responsible for, its credibility is adjusted according to the following formula:

$$C_{mi}^{k+1} = \begin{cases} C_{mi}^k + \frac{(1-C_{mi}^k) \cdot Q_{ij}}{2} & \text{if } |R_{mj} - O_{ij}^{avg}| < s_{mj} \\ C_{mi}^k - \frac{C_{mi}^k \cdot Q_{ij}}{2} & \text{if } |R_{mj} - O_{ij}^{avg}| > s_{mj} \end{cases} \qquad (7)$$

where C_{mi}^k is the credibility of node i after k reports to score manager m, $O_{ij}^a vg$ is the opinion reported by node i, Q_{ij} is the associated quality value, R_{mj} is the aggregated reputation and s_{mj} is the standard deviation of all the reported opinions about node j. Thus, credibility updates take the reported quality value into account. Since an opinion with a smaller quality value does not count as much at the score manager, the change in credibility is proportionately lower. At the highest reported quality value of 1, a reported rating that falls within one standard deviation of the aggregated reputation, increments the credibility of the reporting node by half the amount required for credibility to reach 1. A reported rating outside this region results in the credibility rating dropping to half the previous value.

In this way, if a reporting node (or score manager) is malicious, its credibility rating is gradually reduced when its opinion does not match that of other nodes (or score managers). And a node with a lower credibility value therefore contributes less to the aggregated reputation at the score manager.

3.6   Resource Requirements

Since each node has M score managers associated with it, a transaction between two nodes results in $2M$ messages to the score managers. Similarly, a trust query from a node to the score managers of its potential transaction partner results in M messages to the score managers and M responses. As the number of score managers M does not depend on the number of nodes in the network, the network traffic increases by a constant factor due to the RQC scheme.

Assume that a node transacts with K nodes on average. Since each node acts as a score manager for M nodes and stores the last reported opinion from each of the K transaction partners for these M nodes, the storage requirements for being a score manager are $O(KM)$. Each node also stores its own average opinion for the K nodes it has interacted with, resulting in $O(K)$ storage. Hence the total storage requirements for a node are $O(KM + K) = O(KM)$.

The processing requirements at each node are relatively light. Each transaction (preceded by two trust queries) results in reputation and quality computations at $2M$ score managers (M for each transacting node), followed by weighted averages of the reputation being computed at each transacting node, followed by a single average opinion update at each of the $2M$ score managers.

4   Experimental Results

We simulated the RQC scheme using Pastry [10], a structured overlay network that uses distributed hash tables for routing. We assume that the nodes are always online and full connectivity in the network. We also assume that no messages are dropped and that messages cannot be spoofed or altered in any way. Moreover, nodes do not leave or join the network in the during the simulation. Since Pastry is written in Java, we decided to implement our reputation system in Java as well.

The trust information pertaining to a node i is stored at M score managers that are assigned using a DHT. A hashing function is used to map a persistent node identifier to a point in the key space. The M nodes that are closest to this point in the key space are then used as score managers for that node. A node j can then query all the M score managers in order to compute the reputation of i and decide whether to interact with i.

In the experiments we describe, we simulated a network of 200 nodes unless specified otherwise. In each experiment, 50000 random transactions took place unless specified otherwise (i.e., each node has an average of 50 interactions). Both participants of each interaction are chosen randomly. The default number of score managers storing reputation ratings for each node was 6 unless specified otherwise. The reason we chose a network of this size had to do with the running time for the simulation. We simulated a network of 5000 nodes and $1,000,000$ interaction, but since we were using the Pastry substrate and the entire network was being simulated on just one machine with 1GB of RAM, this took several hours for each run. As the experiment in Figure 5 shows, the impact of the number of nodes was very small on the results of the simulation as long as a sufficient number of interactions took place. Hence, our results should be valid for larger networks as well.

We decided instead, to invest our CPU resources on running each experiment 10 times. In the figures that follow, we plot the average of the results obtained from the 10 experiments, along with a confidence interval of size $\frac{s}{\sqrt{n}}$ where s is the standard deviation and n is the number of samples (10).

We simulate two different kinds of maliciousness. A node can be malicious in the base system, i.e., behave maliciously when interacting with other nodes and/or it may be malicious in the reputation system. In the former case, two good nodes (and two bad nodes) give each other a rating of 1 after interacting whereas if a good and a malicious node interact they both give each other a rating opinion of 0. In the latter case, the node behaves maliciously in its capacity as a score manager and sends incorrect reputation values to requesting nodes.

We also simulate probabilistic maliciousness where a node does not act maliciously all the time and is malicious with a probability p_m which is a parameter to the simulation. Finally, we simulate several scenarios where the number of malicious nodes ranges from 5% of the total population to 90% of the total population.

The performance of our scheme is evaluated as the number of correct decisions made (i.e., interactions with good nodes that went ahead plus interactions with malicious nodes that were avoided) as a proportion of the total number of decisions made. So, an interaction with a malicious node counts against the RQC scheme as much as when an interaction with a good node is prevented due to false ratings. Since, we are interested in the steady state performance of the system, the initial interactions that take place when no information about the reputation of a node can be found are not counted.

4.1 Types of Maliciousness

In Figure 1, we depict the performance of our scheme for different kinds of malicious behavior. The three lines represent the cases when malicious nodes behave maliciously as participants in the base system (i.e., when interacting with other nodes), as participants of the reputation system (i.e., sending false reputations in their capacity as score managers) and as participants in both the base and reputation system.

Malicious behavior in the reputation system is defined as the sending of false reputation values by score managers. A malicious score manager sends a reputation value that is the inverse of the actual (1 minus the actual) reputation value of the nodes it is responsible for. We chose this model of maliciousness as it is the worst possible type of maliciousness from the perspective of a reputation system. Other models of maliciousness, such as sending an arbitrary reputation value in response, do not cause as much harm to the reputation system.

In the case when nodes act maliciously in the base system only, our scheme performs very well till the percentage of malicious nodes reaches 50%. Note that the confidence interval is very large at this point. This shows that our scheme is very sensitive to the proportion of malicious nodes in the system. When the proportion of malicious nodes exceeds half, the dominant ethic of the system becomes that of the malicious nodes. All

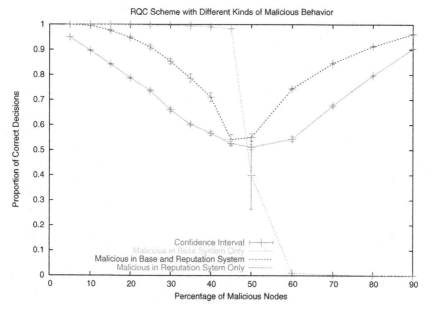

Fig. 1. Performance of the RQC Scheme with Different types of Malicious Nodes

good nodes are branded as malicious and vice-versa resulting in a precipitous drop in performance.

Another striking feature in Figure 1 is that the performance of our scheme goes down and then rises again in both the cases where nodes act maliciously in the reputation system. We assume that malicious nodes are not aware of each others existence and therefore malicious score managers do not treat other malicious nodes any different from other nodes.

When the nodes act maliciously both as participants in the base system and in the reputation system a similar pattern is observed with the worst performance coming when the fraction of malicious nodes is 50%. As the number of malicious nodes increases, a larger proportion of interactions take place between two malicious nodes. These nodes give each other an opinion rating of 1 but when this opinion is reported to the score manager – which itself has a high likelihood of being malicious – it inverts this rating and the reputation of the malicious nodes is correctly reduced to 0. Therefore, a large number of interactions with malicious nodes are avoided, thus improving the performance of our system.

4.2   Comparison with the Aberer-Despotovic Scheme

In this experiment we compared the performance of our scheme against the trust management scheme proposed by Aberer and Despotovic in [5]. We implemented their scheme in Pastry and ran experiments with the same number of nodes and interactions

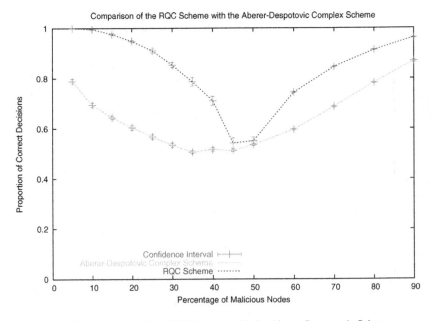

Fig. 2. Comparison of the RQC Scheme with the Aberer-Despotovic Scheme

as in our scheme (200 and 50000 respectively). However, unlike the simulations they performed, we did not make trust assessments at the end of the interaction period but instead made trust assessments before each interaction. We feel this model is closer to reality as nodes would want to know the nature of a node before interacting with it instead of waiting for a large number of interactions to finish. As we can see in Figure 2, our scheme performs consistently better than the Aberer-Despotovic scheme in terms of the proportion of correct decisions made.

Figure 2 compares the performance of the two schemes when there is maliciousness in both the base and the reputation system. While we do not show the corresponding graphs for other types of maliciousness, we would like to note that the RQC scheme also outperforms Aberer-Despotovic for other two models of maliciousness.

4.3 Probabilistic Cheating

In Figure 3 we examine the case when the malicious nodes do not cheat all the time but instead cheat with a certain probability. The three curves correspond to a total of $10\%, 30\%$ and 50% of the nodes being malicious. We see that the proportion of correct decisions is slightly affected by the probability of a node cheating with the RQC scheme generally performing better as nodes cheat more consistently.

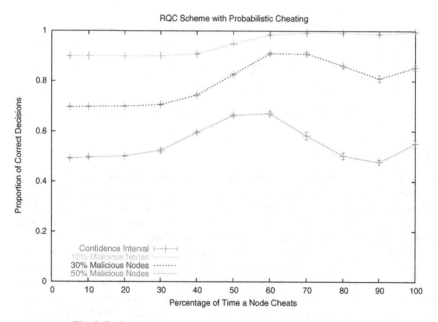

Fig. 3. Performance of the RQC Scheme with Probabilistic Cheating

4.4   Number of Score Managers

In this experiment we study the impact that the number of score managers have on the decision making process. Figure 4 shows the performance of the RQC scheme when 30% of the nodes are malicious and they act maliciously in the base system only. The number of score managers makes no difference to the performance.

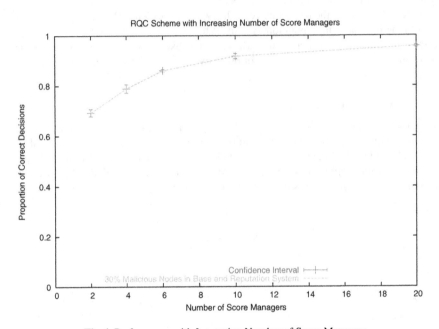

Fig. 4. Performance with Increasing Number of Score Managers

4.5   Number of Nodes

Figure 5 shows how the RQC scheme scales as the number of nodes in the network increases. We ran our simulation with the fraction of malicious nodes at 30%, acting maliciously only in the base system. We ran the experiment with 100, 200, 400, 1000, 2000 and 5000 nodes. There are three curves in the figure corresponding to the total number of interactions being 10, 000, 40, 000 and 160, 000.

We find that an increase in the number of nodes results in a drop in performance if the number of transaction is kept fixed. As the number of nodes increases, the number of interactions per node decreases resulting in less reputation information per node being stored in the system. However, when the number of interaction is increased, the increase in the number of nodes does not have much impact on the performance of the scheme. Hence the scheme scales well with the size of the network if each node takes part in sufficient interactions.

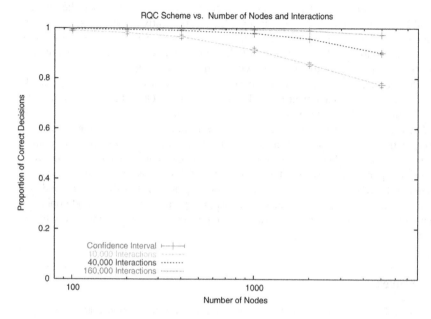

Fig. 5. Performance scaling with Number of Nodes in System

5   Conclusions and Future Work

In this paper, we have presented the **RQC** scheme for trust management in autonomic systems. Our scheme computes Reputation(R), Quality(Q) of Credibility(C) to provide a richer trust management system that lends itself to a wide variety of self-management tasks.

We simulate the RQC scheme using Pastry, a P2P substrate written in Java. However, the RQC scheme can be easily adapted to other environments. The RQC scheme is flexible enough to allow for trust ratings other than 0 or 1. Moreover, the scheme emphasizes consensus as an important indicator of the confidence that can be placed in a rating. This along with the number of interactions forms the basis of the quality of a rating.

The credibility of a node is dependent on the amount by which its opinion of a node (or the reputation it furnishes in case of a score manager) deviates from the mean reputation of the node in question. In [9] the credibility of a node is lowered if its opinion deviates from the mean reputation by more than a constant d. The RQC scheme lowers credibility if the opinion deviates from the mean reputation by more than the sample standard deviation. This does not penalize nodes by lowering their credibility when they report on a node that behaves erratically. The credibility is also predicated on the quality value of the opinion/reputation furnished by a node/score manager. A node should not be penalized as much for an opinion in which it does not have very much confidence.

Our simulation shows that the RQC method performs very well when the number of malicious nodes in the system is under half. It significantly outperforms the Aberer-Despotovic scheme. The scheme continues to work well when the malicious nodes cheat in a probabilistic fashion instead of cheating all the time. Finally, the simulation also shows that the scheme scales well with the number of nodes.

While we apply the RQC scheme to measure node honesty, the scheme also lends itself to other soft-management tasks. For instance, the RQC scheme can be used to measure the reliability or the QoS provided by nodes in the network.

There are still several steps that can be taken to improve the RQC scheme and we are currently working on several enhancements. We are working to incorporate "churn" in our experiments. "Churn" is the phenomenon of nodes joining and leaving a network and is an important component of autonomic and P2P systems. We are also looking at extending the RQC scheme for various network topologies. Finally, we are studying alternative decision-making strategies such as incorporating individual node opinions along with system-wide reputation values when deciding whether to interact with another node.

References

1. Lampe, C., Resnick, P.: Slash(dot) and burn: distributed moderation in a large online conversation space. In: Proceedings of the 2004 conference on Human factors in computing systems, ACM Press (2004) 543–550
2. Resnick, P., Zeckhauser, R., Swanson, J., Lockwood, K.: The value of reputation on ebay: A controlled experiment (2002)
3. Dellarocas, C.: Immunizing online reputation reporting systems against unfair ratings and discriminatory behavior. In: Proceedings of the 2nd ACM conference on Electronic commerce, ACM Press (2000) 150–157
4. Zacharia, G., Moukas, A., Maes, P.: Collaborative reputation mechanisms in electronic marketplaces. In: Proceedings of the Thirty-second Annual Hawaii International Conference on System Sciences-Volume 8, IEEE Computer Society (1999) 8026
5. Aberer, K., Despotovic, Z.: Managing trust in a peer-2-peer information system. In: CIKM. (2001) 310–317
6. Cornelli, F., Damiani, E., di Vimercati, S.D.C., Paraboschi, S., Samarati, P.: Choosing reputable servents in a p2p network. In: Eleventh International World Wide Web Conference, Honolulu, Hawaii (2002)
7. Damiani, E., di Vimercati, S.D.C., Paraboschi, S., Samarati, P.: Managing and sharing servents' reputations in p2p systems. IEEE Transactions on Data and Knowledge Engineering **15** (2003) 840–854
8. Kamvar, S.D., Schlosser, M.T., Garcia-Molina, H.: The eigentrust algorithm for reputation management in p2p networks. In: Proceedings of the twelfth international conference on World Wide Web, ACM Press (2003) 640–651
9. Buchegger, S., Boudec, J.Y.L.: A robust reputation system for p2p and mobile ad-hoc networks. In: Proceedings of the Second Workshop on the Economics of Peer-to-Peer Systems. (2004)
10. Rowstron, A., Druschel, P.: Pastry: Scalable, distributed object location and routing for large-scale peer-to-peer systems. In: IFIP/ACM International Conference on Distributed Systems Platforms (Middleware). (2001) 329–350

E Pluribus Unum*

Deduction, Abduction and Induction, the Reasoning Services for Access Control in Autonomic Communication

Hristo Koshutanski and Fabio Massacci

Dip. di Informatica e Telecomunicazioni - Univ. di Trento
via Sommarive 14 - 38050 Povo di Trento (ITALY)
{hristo, massacci}@dit.unitn.it

Abstract. Autonomic Communication is a new paradigm for dynamic network integration. An Autonomic Network crosses organizational boundaries and is provided by entities that see each other just as business partners. Policy-base network anagement already requires a paradigm shift in the access control mechanism (from identity-based access control to trust management and negotiation), but this is not enough for cross organizational autonomic communication. For many services no partner may guess a priori what credentials will be sent by clients and clients may not know a priori which credentials are required for completing a service requiring the orchestration of many different autonomic nodes.
We propose a logical framework and a Web-Service based implementation for reasoning about access control for Autonomic Communication. Our model is based on interaction and exchange of requests for supplying or declining missing credentials. We identify the formal reasoning services that characterise the problem and sketch their implementation.

1 Introduction

Controlling access to services is a key aspect of networking and the last few years have seen the domination of policy-based access control. Indeed, the paradigm is broader than simple access control, and one may speak of *policy-based network self-management* (e.g. [1] or the IEEE Policy Workshop series). The intuition is that actions of nodes "controlling" the communication are automatically derived from policies. Nodes look at events and requests presented to them, evaluates the rules of their policies and derive actions [1, 2]. Policies can be "simple" `iptables` rules for Linux firewalls (see http://www.netfilter.org/) or complex logical policies expressed in languages such as Ponder [3].

Autonomic Communication adds new challenges: a truly autonomic network is born when nodes are no longer within the boundary of a single enterprise which could deploy its policies on them and guarantee interoperation. Nodes are

* This work is partially funded by EU IST-2001-37004 WASP, EU IST E-NEXT NoE, FIRB RBNE0195K5 ASTRO and FIRB RBAU01P5SS projects.

M. Smirnov (Ed.): WAC 2004, LNCS 3457, pp. 179–190, 2005.

partners that offer services and lightly integrate their efforts into one (hopefully coherent) network. This cross enterprise scenario poses novel security challenges with aspects of both trust management systems and workflow security.

From trust management systems [4, 5, 6] it takes the credential-based view: access to services is offered by autonomic nodes on their own and the decision to grant or deny access must rely on attribute credentials sent by the client. In contrast with these systems, we have a continuous process and assignment of permissions to credentials must look beyond the single access decision.

From workflow access control [7, 8, 9, 10] we borrow all classical problems such as dynamic assignment of roles to users, separation of duties, and assignment of permissions to users according the least privilege principles. In contrast with such schemes, we can no longer assume that the enterprise will assign tasks and roles to users (its employees) in such a way that makes the overall flow possible w.r.t. its security constraints.

Astracting away the details of the policy implementation, we can observe that only one reasoning service is actually used by policy based self-management: *deduction*. Given a policy and a set of additional facts and events, we find out all consequences (actions or obligations) of the policy and the facts, i.e. whether granting the request can be deduced from the policy and the current facts. Policies can be different [11, 6, 12, 8] but the kernel reasoning service is the same.

Autonomic communication needs at least another reasoning service: *abduction* [13]. Loosely speaking, abduction is deduction in reverse: given a policy and a request for access to services, find the credentials/events that would grant access, i.e. a (possibly minimal) set of facts that added to the policy would make the request a logical consequence. Abduction is a core service for the *interactive access control* framework in autonomic communication. In this framework a client may be asked on the fly for additional credentials and the same may disclose them or decline to provide them. We need an interactive control on both the client and server sides whenever the client requires some evidence from the server before disclosing his own credentials.

We might also use *induction* [14]: given a heuristic function to measure the goodness of a rule and some examples of granted and denied requests, invent the access policies covering the positive examples and not the negative ones.

Here, we sketch the reasoning framework for access control for autonomic communication based on interaction for supplying missing credentials or for revoking "wrong" credentials (§2). We identify the reasoning services deduction vs abduction (§4), and induction (§7), and sketch the solution for stateful access control (§5) and mutual negotiation (§6). A running example (§3) makes discussion concrete. A discusssion of future challenges concludes the paper.

2   Access Control with Security Policies

Using Datalog and logic programs for security policy is customary in computer security [11, 6, 12, 8] and our formal model is based on normal logic programs under the stable model semantics [15]. We have predicates for requests, creden-

Role: $R_i \succ$ Role: R_j when role Role: R_i dominates role Role: R_j.
Role: $R_i \succ_{\text{WebServ:}S}$ Role: R_j when role Role: R_i dominates, for service WebServ: S, the role Role: R_j.
assign $(P, \text{WebServ:} S)$ when an access to the service WebServ : S is granted to P. Where P can be either a Role: R or User: U.

(a) Predicates for assignments to Roles and Services

declaration (User: U) it is a statement by the User: U for its identity.
credential (User: U, Role: R) when User: U has a credential activating Role: R.
credentialTask (User: U, WebServ: S) when User: U has the right to access WebServ: S.

(b) Predicates for Credentials

running $(P, \text{WebServ} : S, \text{number:} N)$ when the number: N-th activation of WebServ: S is executed by P.
abort $(P, \text{WebServ:} S, \text{number:} N)$ if the number: N activation of WebServ: S within a workflow aborts.
success $(P, \text{WebServ:} S, \text{number:} N)$ if the number : N-th activation of WebServ : S within a workflow successfully executes.
grant $(P, \text{WebServ:} S, \text{number:} N)$ if the number: N request of WebServ: S has been granted
deny $(P, \text{WebServ:} S, \text{number:} N)$ if the number: N-th request of WebServ: S has been denied.

(b) Predicates for System's History and State

Fig. 1. Predicates used in the model

tials, assignments of users to roles and of roles to services, see Figure 1. They are self explanatory, except for role dominance: a role dominate another if it has more privileges. We have constants for users identifiers, denoted by User: U, for roles, denoted by Role: R, and one for services, denoted by WebServ: S.

Each partner has a *security policy for access control* $\mathcal{P}_A$ and a *security policy for disclosure control* $\mathcal{P}_D$. The former is used for making decision about access to the services offered by the partner. The latter is used to decide the credentials whose need can be potentially disclosed to the client.

We keep a set of *active (unrevoked) credentials* $\mathcal{C}_P$ presented by the client in past requests to other services offered by the same server, and the set of *declined credentials* $\mathcal{C}_N$ compiled from the client's past interactions. To request a service the client submit a set of *presented credentials* $\mathcal{C}_p$, a set of *revoked credentials* $\mathcal{C}_R$ and a *service request* r. We assume that $\mathcal{C}_p$ and $\mathcal{C}_R$ are disjoint. In this context, $\mathcal{C}_N$ is assigned the difference between the missing credentials $\mathcal{C}_M$, the client was asked in the previous interaction, and the ones presented now. For stateful autonomic nodes we'll also need the *history of access to services* $\mathcal{H}$.

3   A Running Example

Let us assume that we have a Planet-Lab shared network between the University of Trento and Fraunhofer institute in Berlin in the context of the E-NEXT network, and that there are three main access types to the resources: *read* – access to data residing on the Planet-Lab machines; *run* – access to data and possibility to run processes on the machines; and *configure* – including the previous two types of accesses plus the possibility of configuring network services on the machines.

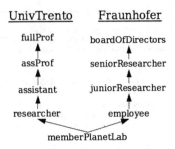

Fig. 2. Joint Hierarchy Model

All Planet-Lab credentials (certificates) are signed and issued by trusted authorities and the crypto validation is performed before the actual access control process. In other words, a preprocessing step validates and transforms the certificates into a form suitable for the formal model – credential (User : U, Role : R).

Fig. 2 shows the role hierarchy, where higher the role in the hierarchy, more powerful it is. A role dominates another role if it is higher in the hierarchy and there is a direct path between them. Fig. 3 shows the access and disclosure policies. authNetwork $(IP, DomainName)$ is domain specific: the first argument is the IP address of the authorized network endpoint (the client's machine) and the second one the domain where the IP address comes from.

Example 1. Rules (1,2) give access to the shared network content to everybody from UniTrento and Fraunhofer, regardless of IP and role. For rules (6,7), if a user has got a disk access and is a researcher at UniTrento or junior researcher at Fraunhofer, it has additional rights. Rules (10,11) give full access from anywhere only to members of the board of directors and to full professors.

Example 2. Rule (5) relaxes the previous two and allows access from any place of the institutions provided users declare their ID and present some role-position certificate of their organization or at least a Planet-Lab membership credential.

Example 3. Rules (1,2) in the disclosure policy show the need for the client to declare its ID if the same comes from an authorized network of the respective organizations; rule (3) discloses the need for Planet-Lab membership credential if the client has already declared its ID; and rule (4) discloses (upgrades) the need of a higher role-position credential.

4   Deduction vs Abduction

The basic reasoning service for policy-based approches is deduction:

Definition 1 (Logical Consequence and Consistency). *We use the symbol* $P \models L$, *where* P *is a policy and* L *is either a credential or a service request, to specify that* L *is a logical consequence of a policy* P. P *is consistent* $(P \not\models \bot)$ *if there is a model for* P.

Access Policy:

(1) assign $(*, request(read)) \leftarrow$ authNetwork $(*, *.unitn.it)$.

(2) assign $(*, request(read)) \leftarrow$ authNetwork $(*, *.fraunhofer.de)$.

(3) assign $(*, request(execute)) \leftarrow$ authNetwork $(193.168.205.*, *.unitn.it)$.

(4) assign $(*, request(execute)) \leftarrow$ authNetwork $(198.162.45.*, *.fraunhofer.de)$.

(5) assign $(User, request(execute)) \leftarrow$ assign $(User, request(read))$, declaration $(User)$, credential $(User, Role)$, $Role \succeq memberPlanetLab$.

(6) assign $(User, request(addService)) \leftarrow$ assign $(User, request(execute))$, declaration $(User)$, credential $(User, Role)$, $Role \succeq researcher$.

(7) assign $(User, request(addService)) \leftarrow$ assign $(User, request(execute))$, declaration $(User)$, credential $(User, Role)$, $Role \succeq juniorResearcher$.

(8) assign $(User, request(addService)) \leftarrow$ authNetwork $(*, *.it)$, declaration $(User)$, credential $(User, Role)$, $Role \succeq assProf$.

(9) assign $(User, request(addService)) \leftarrow$ authNetwork $(*, *.de)$, declaration $(User)$, credential $(User, Role)$, $Role \succeq seniorResearcher$.

(10) assign $(User, request(addService)) \leftarrow$ authNetwork $(*, *)$, declaration $(User)$, credential $(User, Role)$, $Role \succeq fullProf$.

(11) assign $(User, request(addService)) \leftarrow$ authNetwork $(*, *)$, declaration $(User)$, credential $(User, Role)$, $Role \succeq boardOfDirectors$.

Release Policy:

(1) declaration $(User) \leftarrow$ authNetwork $(*, *.unitn.it)$.

(2) declaration $(User) \leftarrow$ authNetwork $(*, *.fraunhofer.de)$.

(3) credential $(memberPlanetLab, User) \leftarrow$ declaration $(User)$.

(4) credential $(RoleX, User) \leftarrow$ credential $(RoleY, User)$, $RoleX \succ RoleY$.

Fig. 3. Proxy Access and Release Policies for the Online Library

This reasoning service is used in most logical formalizations [16]: if the request r is a consequence of the policy and the credentials (i.e. $\mathcal{P}_A \cup \mathcal{C}_p \models r$), then access is granted otherwise it is denied.

Example 4. A request coming from *dottorati.dit.unitn.it* with IP *193.168.205.11* for access to a fellowship application form on the subnet is granted by rule (3).

The next service is abduction: given a policy and a request, find the credentials that added to the policy would allow to grant the request.

Definition 2 (Abduction). *The abductive solution over a policy P, a set of predicates (credentials) H (with a partial order $\prec$ over subsets of H) and a ground literal L is a set of ground atoms E such that: (i) $E \subseteq H$, (ii) $P \cup E \models L$, (iii) $P \cup E \not\models \perp$, (iv) any set $E' \prec E$ does not satisfy all conditions above.*

Traditional p.o.s are subset containment or set cardinality. Other solutions are possible with orderings over predicates.

This reasoning service is used in the overall interactive access control algorithm shown in Fig. 4. Initially the client will send a set of client's credentials $\mathcal{C}_p$ and a service request r. Then we update client's profile, i.e. declined and active credentials and check whether the active credentials unlock r according to $\mathcal{P}_A$. In the case of denial, we compute all credentials disclosable from $\mathcal{C}_p$ according to $\mathcal{P}_D$ and from the resulting set remove all $\mathcal{C}_N$. Then we compute all possible subsets of $\mathcal{C}_D$ that are consistent with the access policy $\mathcal{P}_A$ and, at the same

---

Global vars: $\mathcal{C_N}, \mathcal{C_P}$;
Internal input: $\mathcal{P_A}, \mathcal{P_D}$;
Output: grant/deny/ask($\mathcal{C_M}$);

1. **client's input:** $\mathcal{C}_p$ and r,
2. update $\mathcal{C_N} = (\mathcal{C_N} \cup \mathcal{C_M}) \setminus \mathcal{C}_p$, where $\mathcal{C_M}$ is from the last interaction,
3. update $\mathcal{C_P} = \mathcal{C_P} \cup \mathcal{C}_p$,
4. verify that the request r is a security consequence of the policy access $\mathcal{P_A}$ and presented credentials $\mathcal{C_P}$, namely $\mathcal{P_A} \cup \mathcal{C_P} \models r$ and $\mathcal{P_A} \cup \mathcal{C_P} \not\models \bot$
5. <u>if</u> the check succeeds <u>then</u> return **grant** <u>else</u>
   (a) compute the set of *disclosable credentials* $\mathcal{C_D}$ as
       $\mathcal{C_D} = \{c \mid c \text{ credential that } \mathcal{P_D} \cup \mathcal{C_P} \models c\} \setminus (\mathcal{C_N} \cup \mathcal{C_P})$,
   (b) use abduction to find a minimal set of missing credentials $\mathcal{C_M} \subseteq \mathcal{C_D}$ such that both $\mathcal{P_A} \cup \mathcal{C_P} \cup \mathcal{C_M} \models r$ and $\mathcal{P_A} \cup \mathcal{C_P} \cup \mathcal{C_M} \not\models \bot$,
   (c) <u>if</u> no set $\mathcal{C_M}$ exists <u>then</u> return **deny** <u>else</u>
   (d) return **ask($\mathcal{C_M}$)** and iterate.

Fig. 4. Interactive Access Control Algorithm

time, grant r. Out of all these sets (if any) the algorithm selects the minimal one. We point out that the minimality criterion could be different for different contexts (see [17] for some examples).

Remark 1. Using declined credentials is essential to avoid loops in the process and to guarantee the success of interaction in presence of disjunctive information.

For example suppose we have alternatives in the partner's policy (e.g., "present either a VISA or a Mastercard or an American Express card"). An arbitrary alternative can be selected by the abduction algorithm and on the next inter-action step (if the client has declined the credential) the abduction algorithm is informed that the previous solution was not accepted.

Example 5. Assuming the access and release policies in Figure 3, let us play the following scenario. A senior researcher at Fraunhofer institute FOKUS wants to reconfigure an online service for paper submissions, of a workshop. The service is part of a big management system hosted at the University of Trento's network that is part of Planet-Lab. So, for doing that, at the time of access, he presents his employee membership token, issued by a Fraunhofer certificate authority, presuming that it is enough as a potential customer.

Formaly speaking, the request comes from a domain *fokus.fraunhofer.de* with credential for Role : *employee* together with a declaration for a user ID, *John Milburk*. According to the access policy the credentials are not enough to get full access and so the request would be denied.

Then, following the algorithm in Figure 4, it is computed the set of disclosable credentials from the disclosure policy and the user's available credentials, and the minimal set of credentials, out of those, that satisfies the request. The resulting set is {credential (User: $JohnMilburk$, Role: $juniorResearcher$)}. Then the need for this credential is return back to the user.

Example 6. On the next interaction step, because the user is a senior researcher, the same declines to present the requested credential as just returning the same query with no presented credentials.

So, the algorithm updates the user's session profile and the outcome is the need for credential credential (User: $JohnMilburk$, Role: $seniorResearcher$).

5   Stateful AC: Missing and Excessing Credentials

What happens if access to services is determined also by the history of past executions? For instance in the example by Atluri and Bertino [8–pag.67] a branch manager of a bank clearing a cheque cannot be the same member of staff who has emitted the cheque. So, if we have no memory of past credentials then it is impossible to enforce any security policy for separation of duties on the application workflow. The problems are the following:

— the request may be inconsistent with some role used by the client in the past;
— the new set of credential may be inconsistent with requirements such as separation of duties;
— in contrast to intra-enterprise workflow systems [8], the partner offering the service has no way to assign to the client the right set of credentials which would be consisted with his future requests (because he cannot assign him future tasks).

So, we must have some roll-back procedure by which, if the user has by chance sent the "wrong" credentials, he can revoke them.

Our interactive access control solution for stateful services and applications is shown in Figure 5.

The logical explanation of the algorithm is the following. Initially when a client requests a specific service the authorization mechanism creates a new session with global variables declined credentials $\mathcal{C}_\mathcal{N}$, not revoked credentials $\mathcal{C}_\mathcal{U}$, missing credentials $\mathcal{C}_\mathcal{M}$ and excessing credentials $\mathcal{C}_\mathcal{E}$ set up to empty sets. Then once the session is started, internally, the algorithm loads the policies for access and disclosure control $\mathcal{P}_\mathcal{A}$ and $\mathcal{P}_\mathcal{D}$ together with the two external sets history of execusion $\mathcal{H}$ and client's active credentials $\mathcal{C}_\mathcal{P}$.

Following that, the first step in the algorithm is to get the client's input as sets of currently presented credentials $\mathcal{C}_p$, the revoked ones $\mathcal{C}_r$ and the service request r. Then the set of active credentials $\mathcal{C}_\mathcal{P}$ is updated as removing the set $\mathcal{C}_r$ from it and then adding the set of currently presented credentials (rf. step 2). Then in step 3 declined credentials $\mathcal{C}_\mathcal{N}$ are updated as credentials the client was asked in the last interactions minus the ones that he has currently presented. Analogously, in step 4, not revoked credentials $\mathcal{C}_\mathcal{U}$ are updated as the excessing

---

Global vars: $C_N, C_U, C_M, C_{\mathcal{E}}$;   Initially $C_N = C_U = C_M = C_{\mathcal{E}} = \emptyset$;
Internal input: $\mathcal{P}_A, \mathcal{P}_D, \mathcal{H}, C_P$;
Output: grant/deny/$< ask(C_M), revoke(C_{\mathcal{E}}) >$;

1. **client's input:** C_p, C_r and r,
2. update $C_P = (C_P \setminus C_r) \cup C_p$,
3. update $C_N = (C_N \cup C_M) \setminus C_p$, where C_M is from the last interaction,
4. update $C_U = (C_U \cup C_{\mathcal{E}}) \setminus C_r$, where $C_{\mathcal{E}}$ is from the last interaction,
5. Set up $C_M = C_{\mathcal{E}} = \emptyset$,
6. verify whether the request r is a security consequence of the policy access $\mathcal{P}_A$ and presented credentials C_P, namely $\mathcal{P}_A \cup \mathcal{H} \cup C_P \models r$ and $\mathcal{P}_A \cup \mathcal{H} \cup C_P \not\models \bot$,
7. <u>if</u> the check succeeds <u>then</u> return **grant** <u>else</u>
   (a) compute the set of *disclosable credentials* $C_D = \{c \mid \mathcal{P}_D \cup C_P \models c\} \setminus (C_N \cup C_P)$,
   (b) use abduction to find a minimal set of *missing credentials* $C_M \subseteq C_D$ such that both $\mathcal{P}_A \cup \mathcal{H} \cup C_P \cup C_M \models r$ and $\mathcal{P}_A \cup \mathcal{H} \cup C_P \cup C_M \not\models \bot$,
   (c) <u>if</u> a set C_M exists <u>then</u> return $< ask(C_M), revoke(C_{\mathcal{E}}) >$ <u>else</u>
      i. use abduction to find a minimal set of missing credentials $C_M \subseteq (C_D \cup C_P)$ such that $\mathcal{P}_A \cup \mathcal{H} \cup C_M \models r$, $\mathcal{P}_A \cup \mathcal{H} \cup C_M \not\models \bot$ and $C_U \cap (C_P \setminus C_M) = \emptyset$,
      ii. <u>if</u> no set C_M exists <u>then</u> return **deny** <u>else</u>
      iii. compute $C_{\mathcal{E}} = C_P \setminus C_M$ and $C_M = C_M \setminus C_P$,
      iv. return $< ask(C_M), revoke(C_{\mathcal{E}}) >$ and iterate.

Fig. 5. Interactive Access Control Algorithm for Stateful Autonomic Services

credentials asked in the last interaction minus the ones currently revoked. Step 5 prepares the two sets C_M and $C_{\mathcal{E}}$ for the interaction output.

Steps 6, 7, 7a, 7b and 7c have the same explanation as the respective ones in Figure 4. If a set of missing credentials was not found in step 7b then we run the abduction process again (step 7(c)i) but over the extended set of disclosable credentials and active credentials $C_D \cup C_P$ searching for a solution for r that preserves consistency in $\mathcal{P}_A$ and unlocks r. The last requirement in the step is used to filter out those solutions that have been partially refused to be revoked.

Step 7(c)i indicates that if a set C_M exists then definitely there are "wrong" credentials among those in C_P that ban the client to get a solution for r (in step 7b). If no such set then the client is denied because he does not have enough privileges to disclose more credentials to obtain the service r (step 7(c)ii).

Step 7(c)iii computes the sets of excessing and missing credentials $C_{\mathcal{E}}$ and C_M. The motivation behind $C_{\mathcal{E}}$ is that the set difference of active credentials minus just computed C_M certainly contains the credentials that ban the client to get a solution for r.

At this point there two main issues concerning the set $C_{\mathcal{E}}$: (i) the system may restart from scratch asking the client to revoke all his active credentials, i.e. $C_{\mathcal{E}} = C_P$, (ii) the system may ask the client to present credentials that have been already asked for revocation in past interactions.

Remark 2. Step 7(c)iii looks the opposite of abduction: rather than adding new information to derive more things (the request), we drop information to derive less things (the inconsistency). One can show that the two tasks are equivalent.

6   Life Is Complicated: Two-Party Negotiation

So far deduction helps us to infer whether a service request is granted by the partner's access policy and the client's set of credentials. In the case of failure, abduction infers what is missing so that the client can still get the desired service.

Example 7. When the senior researcher received the counter request to present his *seniorResearcher* certificate in order to get access he may not want to reveal his role if he is not sure that he talks with a University of Trento's server.

We should allow him to request the system to show a certificate. The system, in its turn, may have policy saying that such certificates are disclosed only to entities coming from an authorized network, e.g., authNetwork $(*, *.fraunhofer.de)$.

The next step is how to establish and automate a two-party negotiation process using the inference capabilities on both sides. For that purpose we need to extend each of the party's policies:

– a policy for access to *own* resources $\mathcal{P}_{A\mathcal{R}}$ on the basis of *foreign* credentials,
– a policy for access to *own* credentials $\mathcal{P}_{AC}$ on the basis of *foreign* credentials,
– a policy for the disclosure of the need of missing *foreign credentials* $\mathcal{P}_{\mathcal{D}}$.

Client and the server just have to run the same negotiation protocol:

1. The client, Alice, sends a service request r and (optionally) a set of credentials $\mathcal{C}_p$ to the server, Bob.
2. Then Bob looks at r and if it is a request for a service he calls the interactive access control algorithm in Figure 4 with his policies for access and disclosure of resources $< \mathcal{P}_{A\mathcal{R}}, \mathcal{P}_{\mathcal{D}} >$.
3. If r is a request for a credential then he calls the same algorithm with his respective policies for access and disclosure of credentials $< \mathcal{P}_{AC}, \mathcal{P}_{\mathcal{D}} >$.
4. In the case of computed missing credentials $\mathcal{C}_{\mathcal{M}}$, he transforms that into counter-requests for credentials and waits until receives all responses. At this point Bob acts as a client, requesting Alice the set of credentials $\mathcal{C}_{\mathcal{M}}$. Alice will run the same protocol swapping roles.
5. When Bob's main process receives all responses it checks whether the missing credentials have been supplied by Alice.
6. If $\mathcal{C}_{\mathcal{M}}$ was not reached, Bob restarts the loop and consults the interactive access control algorithm for a new decision.
7. When a final decision is taken, the response (grant/deny) is sent to Alice.

The protocol can be run on both sides so that they can communicate and negotiate the missing credentials until enough trust is established and the service is granted or the negotiation failed and the process is terminated.

7   Induction: Finding the Rules

The work for inductive logic programming [14] has been most evolved in the field of machine learning. Inductive logic programming systems (ILP) construct concept definitions from examples and a logical domain theory.

Induction may be an extremnely valuable tool for autonomic nodes, because complete and consistent access policies may be difficult to write. So it might be weel the case that a node has only a partial policy, and some additional set of examples of access that one desired to permit or forbid. Then the node should be able, by generalizing from the examples to derive a policy that matched the given examples and is also asble to answer other similar queries.

So an autonomic node could be provided with background police P_B, some sample granted requests for services R^+ and denied requests R^- and as a result it should be able to constructs a tentative access policy P_{HA}. Here R^+, R^- are sets of ground facts and P_B and P_{HA} are logic programs. The conditions for construction of P_{HA} are:

Necessity: $P_B \not\models R^+$,
Sufficiency: $P_B \wedge P_{HA} \models E^+$,
Weak consistency: $P_B \wedge P_{HA} \not\models \bot$,
Strong consistency: $P_B \wedge P_{HA} \wedge E^- \not\models \bot$.

A number of algorithms can be used for determining the construction of P_{HA} based on ILP (see e.g. [14]) and the identification of the most appropriate for autonomic communication policies is the subject of future work.

8   Implementation

We have implemented a system for access control for abduction and deduction using protocols over web services, a front-end to a state-of-the-art inference engine and integrated it with a system for PMI (privilege management infrastructure).

For our implementation, Collaxa[1] is used as a main manager of Web Services Business Processes (on the AuthorizationServer side).

PolicyEvaluator is a Java module that acts as a wrapper for the DLV system[2] (a disjunctive datalog system with negations and constraints) and implements our interactive algorithm for stateless autonomic nodes (Fig. 4). For deductive computations we use the disjunctive datalog front-end (the default one) while for abductive computations, the diagnosis front-end.

The current system processes credentials at an high level: defines what can be inferred and what is missing from a partner's access policy and a user's set of credentials. For the actual distributed management of credentials at lower levels (namely actual cryptographic verification of credentials) we decided to use

---

[1] Collaxa BPEL Server (v2.0 rc3) – www.collaxa.com
[2] DLV System (r. 2003-05-16) – www.dlvsystem.com

PERMIS infrastructure [18] because it incorporates and deals entirely with X.509 Identity and Attribute Certificates. It allows for creating, allocating, storing and validating such certificates. Since PERMIS conforms to well-defined standards we can easily interoperate with the other entities (partners) in the network.

9   The Challenges Ahead

So far we have presented a logical framework and a proof-of-concept implementation for reasoning about access control for autonomic communication based on interaction for supplying missing credentials or for revoking "wrong" credentials. We have discussed the different formal reasoning services – deduction, abduction, and induction, with a special emphasis of the first two. We have also show how the model can deal with stateful access to services and two party negotiation.

Yet, a number of major challenges remains:

Complexity Characterization: abduction engines such as DLV are rather effective but unfortunately general algorithms for abduction are inefficient[3]. Our problems are at the same time more specialized (e.g. credentials are occurring only positively in the rules) and more general (we have hierarchies of roles so subset or cardinality minimality does not really apply). So capturing the exact computational complexity of the problem may be far from trivial.

Approximation vs Language Restriction: even if the problem is hard in the general case, we might have suitable syntactic restrictions that allow for a polynomial evaluation. In other cases, we may be able to find out anytime algorithms that gives an approximate answer (not really the minimal one but close to it).

Reputation Management: so far we have assumed that declining or presenting a credential has no impact on the reputation of nodes. Research on algorithms and logics for secure reputation is still in the early stage but its integration with interactive access control might have a significant impact.

Negotiatation Strategy Analysis: which is the impact of the negotiation strategy on the effectiveness, completeness, privacy protection, immunity from DoS attacks of interactive access control? So far only the completeness of the procedure is settled and more sophisticated strategies, taking into account the value of credentials that are disclosed could lead to many interesting results relevant for the practical deployment of the framework.

Policy Compilation: this is likely the topic with major impact on industry. All policies (either in networking or security) are either interpreted or hardwired in the application. In contrast, we would need a way to "compile" the policy and the policy enforcement engine into machine languages so that autonomic nodes can quickly react to the requests and yet gives us the flexibility of policies: update a policy simply means recompiling and redeploying.

---

[3] They lay at the secon level of the polynomial hierarchy, i.e. harder than NP.

References

1. Sloman, M., Lupu, E.: Policy specification for programmable networks. In: 1st Inter. Working Conference on Active Networks, Springer-Verlag (1999) 73–84
2. Smirnov, M.: Rule-based systems security model. In: Proceedings of the Second International Workshop on Mathematical Methods, Models, and Architectures for Computer Network Security (MMM-ACNS), Springer (2003) 135–146
3. Damianou, N., Dulay, N., Lupu, E., Sloman, M.: The Ponder policy specification language. In: Proceedings of the International Workshop on Policies for Distributed Systems and Networks (POLICY), Springer-Verlag (2001) 18–38
4. Weeks, S.: Understanding trust management systems. In: IEEE Symposium on Security and Privacy (SS&P), IEEE Press (2001)
5. Ellison, C., Frantz, B., Lampson, B., Rivest, R., Thomas, B.M., Ylonen, T.: SPKI Certificate Theory. (1999) IETF RFC 2693.
6. Li, N., Grosof, B.N., Feigenbaum, J.: Delegation logic: A logic-based approach to distributed authorization. ACM TISSEC 6 (2003) 128–171
7. Atluri, V., Chun, S.A., Mazzoleni, P.: A Chinese wall security model for decentralized workflow systems. In: Proceedings of the 8th ACM CCS. (2001) 48–57
8. Bertino, E., Ferrari, E., Atluri, V.: The specification and enforcement of authorization constraints in workflow management systems. ACM TISSEC 2 (1999) 65–104
9. Georgakopoulos, D., Hornick, M.F., Sheth, A.P.: An overview of workflow management: From process modeling to workflow automation infrastructure. Distributed and Parallel Databases 3 (1995) 119–153
10. Kang, M.H., Park, J.S., Froscher, J.N.: Access control mechanisms for inter-organizational workflow. In: 6th ACM SACMAT. (2001) 66–74
11. Bertino, E., Catania, B., Ferrari, E., Perlasca, P.: A logical framework for reasoning about access control models. In: 6th ACM SACMAT. (2001) 41–52
12. Bonatti, P., Samarati, P.: A unified framework for regulating access and information release on the web. Journal of Computer Security 10 (2002) 241–272
13. Shanahan, M.: Prediction is deduction but explanation is abduction. In: Proceedings of IJCAI '89, Morgan Kaufmann (1989) 1055–1060
14. Muggleton, S., De Raedt, L.: Inductive logic programming: Theory and methods. JLP 19/20 (1994) 629–679
15. Apt, K.: Logic programming. In van Leeuwen, J., ed.: Handbook of Theoretical Computer Science. Elsevier (1990)
16. De Capitani di Vimercati, S., Samarati, P.: Access control: Policies, models, and mechanism. In Focardi, R., Gorrieri, F., eds.: Foundations of Security Analysis and Design - Tutorial Lectures. Volume 2171 of LNCS. Springer Verlag Press (2001)
17. Koshutanski, H., Massacci, F.: Interactive access control for Web Services. In: 19th IFIP Information Security Conference (SEC), Kluwer Press (2004) 151–166
18. Chadwick, D.W., Otenko, A.: The PERMIS X.509 role-based privilege management infrastructure. In: Seventh ACM SACMAT. (2002) 135–140

A Metabolic Approach to Protocol Resilience

Christian Tschudin[1] and Lidia Yamamoto[2]

[1] Computer Science Department, University of Basel,
Bernoullistrasse 16, CH-4056 Basel, Switzerland
Christian.Tschudin@unibas.ch
[2] Hitachi Europe, Sophia Antipolis Laboratory,
1503 Route des Dolines, F-06650 Valbonne, France
Lidia.Yamamoto@hitachi-eu.com

Abstract. The goal of this research is to create robust execution circuits for communication software which can distribute over a network and which continues to provide its service despite parts of the implementation being knocked out. Like packets that can be lost (which can be recovered by the appropriate protocols) we envisage an environment where parts of a protocol's execution can be lost. The remaining implementation elements should continue to operate and be able to recover by themselves for restoring full services again. Based on a chemical execution model, we show a few initial examples of packet processing functions that are robust against the knock-out of any single instruction. These examples illustrate how the model can be applied to implement resilient communication protocols, to which we add regulatory signals that can be used to steer the protocols' code basis.

Keywords: resilient communication software, autonomic communication, bio-inspired networking, active networking, Fraglets.

1 Introduction

Autonomic Communication [1] is a long-term research initiative aimed at the study of the self-organization of network elements, toward their autonomous behavior and automated evolvability. Autonomic networks must be self-managing, which includes self-monitoring and self-healing, among other self-* properties. Several areas are concerned, including security, trust, stability, resilience, control, programmability, behavior composition, and context awareness.

Two important and complementary goals of autonomic communication are resilience and self-healing capacity: resilience against internal failures and misbehavior, and self-healing ability to recover from such abnormal conditions.

Resilience and self-healing ability are essential properties of a self-organizing network, where functional and coherent protocol structures must emerge out of basic protocol submodules. The system must be able to detect and replace misbehaving software at run time, while continuing to provide the service, although perhaps less efficiently during the transitory repair phase.

M. Smirnov (Ed.): WAC 2004, LNCS 3457, pp. 191–206, 2005.

In most commonly encountered current computer systems there is always a risk of full service disruption due to buggy or malicious code. The underlying software systems are usually not robust to misbehaving code, and are unable to autonomously resume their normal behavior after such misbehaving code has been installed. This fragility stems from the implicit assumption that all code should be well-behaving, predictable and correct. This assumption is unrealistic, as it can be observed daily in the form of disruptive software bugs, viruses, worms, and attacks of various kinds.

The current methodology for the design of protocols is not better. It relies on the strong assumption of full reliability of the executing components. The fragility of this approach is easy to demonstrate: remove the processor, remove the software module, remove a procedure, or even a single instruction. All these will with high probability lead to the protocol failing to provide the intended communication service.

We concentrate on the last aspect, i.e. the impact of removing a single instruction, or small code fragment, and formulate a first goal: a protocol implementation should be robust enough such that perturbing any single instruction does not lead to wrong protocol behavior. Note that the definition of a single instruction or code fragment depends on the instruction set used, and will be explained in Section 4.

The second goal is to be able to detect and correct wrong protocol behavior, provided that the amount of error is below a threshold. Such self-healing ability is essential to maintain the first goal (resilience) during a potentially long period of operation. Otherwise errors could accumulate, leading sooner or later to service disruption.

This is analogous to the desired properties of forward error correcting codes where a single bit flip still permits to recover and correct a message, or a reliable transfer protocol where message retransmission can be requested, which in our case would correspond to rectifying an incomplete protocol software.

In other words: Like the basically unreliable transmission of messages where messages can be lost, reordered etc., we assume that some subset of the execution paths of the protocol software are executed in an unreliable way. We then ask whether communication software can be written such that it is able to recover itself in such circumstances.

If this can be achieved, it means that altering a single instruction will not do any harm to the protocol in question. A consequence of this is that it becomes in theory possible to disperse the code of this resilient protocol, such that each atomic instruction would be carried out by a different processor. Register values could be shipped in packets between the nodes. Since the protocol is robust against the loss of a single instruction, it is now robust against the crash of any of the processors involved. In practice such partitioning would not occur on a single instruction basis but at the level of modules or code compartments. In this case we would like that any compartment crash be tolerable, while keeping instruction-level robustness inside the compartment. Robustness can be examined at different levels.

The applications of resilient and self-healing protocols are numerous: they would enable safe automated installation of new protocols, protocol upgrade, run-time customization of protocols to adapt to different network situations, distributed protocol implementations in sensor networks, spray computers [2], support for ambient intelligence and other networks of small devices, the dynamic placing of middle-box elements such as proxies and caches, and so on.

As a first approach to the problem we take inspiration from metabolic pathways in cells. These chemical processes are highly interlocked and surprisingly robust. This is of major interest to the pharmaceutical industry that is faced with the problem of identifying the multiple change points in a metabolic pathway in order to alter a cell's production levels (e.g. reproduction of a virus, cancer cell, etc.), where a single inhibition point is in general hard to find.

We build upon previous work on fraglets [3] as it permits to easily demonstrate a simple example of a robust piece of software. The fraglet model comprises a unified code/data format and execution engine inspired by metabolic networks in molecular biology.

The rest of this paper is organized as follows: Section 2 discusses related techniques including self-testing software, fault tolerant systems, resilience in today's protocols, and so on. Section 3 describes parallel execution frameworks for communication software and introduces the fraglet model and instruction set. Section 4 gives a few initial hints on how resilient protocols can be constructed such that the loss of a (fraglet) rule has no impact on the service but leaves some traces behind which can be used to trigger a self-healing process. Section 5 presents our current conclusions and ideas to motivate a new branch of autonomic communication dedicated to resilient and self-healing code for autonomic network protocols.

2   Related Work

2.1   Protocol Robustness Today

Resilience is a key requirement for any network. At the hardware level it is common to have redundant links, and redundant parts in core-network routers. At the software level, most current protocols and network services incorporate some form of robustness. We give a few examples below.

Robustness to link or node failure is generally achieved by rerouting traffic to alternative paths. In OSPF (Open Shortest Path First), node or link failures are detected via link advertisement messages, and new routes are recomputed accordingly. In BGP (Border Gateway Protocol), as well as in MPLS (Multi-Protocol Label Switching) route restoration is achieved via backup paths, such that service can be preserved during failure of the main path.

At the transport layer, TCP recovers from packet loss, duplicate packets, and congested paths, via an integrated retransmission and congestion control mechanism. Other adaptive transmission schemes are able to recover from similar disturbances by using mechanisms adapted to the nature of the traffic they transport, such as real-time streaming, or loss-tolerant voice/video.

The DNS (Domain Name Server) resolution service is a key pillar of the Internet. Resilience is a paramount concern, and is achieved through redundant servers.

Resilient Overlay Networks (RON) [4] are application-layer overlays on top of the current Internet. They seek to improve end-to-end reliability and performance by dynamically avoiding overloaded or faulty paths. RON nodes monitor path quality in order to detect and select higher quality paths. This makes it possible for the end systems to self-organize into more robust topologies that could not otherwise be offered by the standard Internet routing mechanisms.

These classic robustness protocols focus on external events such as node or link failure, link errors or congestion and are not robust to failure of the protocol implementation itself, except if the failure of an implementation on a node corresponds to the failing of the full node.

2.2   Fault Tolerance in Distributed Systems

The complexity of distributed systems and their dependence upon the underlying hardware and network infrastructure expose them to several possible faults. Fault tolerance must therefore be an inherent part of their design, so that they can keep delivering the intended services with acceptable levels of performance and safety, even in the presence of failures.

There are multiple techniques for fault tolerance in distributed systems [5, 6, 7, 8], but most of them use variants or combinations of two main building blocks: state persistence and redundancy.

State persistence can be achieved via checkpointing and/or logging. A checkpoint is a snapshot of the process state at a given execution point. A log contains incremental state changes. Both can be used to resume execution after a failure, at the latest stage for which state has been consistently saved. This can diminish the impact of the failure.

Redundancy can take the form of multiple copies of the same process running on different machines, or multiple versions of a software component which implement the same functionality in different ways. For example, if failure of a version is detected, one can substitute the bad version with another one and try again.

A typical way to implement a fault-tolerant service is by replicating the servers at several independent locations and coordinating the updates so that at least one of the servers is available. When redundant processes (either via simple replication of code or via multiple versions) are used to achieve a result, a voting scheme is usually employed to decide on the output that must be actually produced.

Reconfiguration can be performed to recover from faulty processors by replicating the process to an operational processor, or to replace a malfunctioning version with a correct one. A model to dynamically reconfigure software in distributed systems is presented in [9]. It analyzes the dependencies among the different processes in the system to determine the impact of a given reconfiguration operation. The model has been used to build a fault-tolerant environment

based on ARMOR processes (Adaptive Reconfigurable Mobile Objects of Reliability). However the model itself does not advise on which kind of reconfiguration
should be performed under which circumstances. It leaves this task to system
administrators or possibly other software components with the ability to make
such decisions. Moreover, if the checkpointing or replication logic becomes cor-
rupted, fault tolerance will not be achievable anymore. We call this an extrinsic
approach where fault tolerance logic is incorporated into a system as an add
on. Our interest is in an intrinsic solution where the robustness is part of the
(protocol) software itself.

2.3   Self-testing and Self-correcting Software

The usual methodologies for software testing and debugging rely on running the
program under a controlled environment using a subset of possible inputs. The corresponding outputs must be known beforehand, and in most cases bugs arise after
the software is deployed, due to untested combinations of input data or events.

Self-testing and self-correcting programs [10, 11] have been proposed to improve the reliability of software systems. Given a program P, a self-testing program for P is another, simpler program designed to make calls to P on a number
of inputs, and to check whether the corresponding outputs are correct. A self-
corrector for the same program P attempts to return the correct output for each
input value, even in the presence of abnormal behavior of P, provided that the
probability of P producing the wrong output is small enough.

While these techniques can in principle be useful for software debugging and
testing, and to improve software reliability at run time, in practice especially the
self-correcting functions can be hard to design, since they are very specific for
each program. Another shortcoming of self-correcting functions is that they do
not correct the program itself, but just attempt to correct wrong outputs.

In [12] the authors extend the notion of self-testing/correcting functions to
distributed processing or protocols. Their approach has the same advantages
and limitations of [10, 11], but applied to protocols in distributed systems, which
makes the testing and correcting algorithms more complex.

Again, self-testing or correction is an extrinsic approach that works on a given
program: it does not apply to self-testing and correcting its own operation.

2.4   Unfaithful Mobile Code Execution and Its Detection

The "malicious host problem" refers to an execution environment which actively
tries to distort a program's execution or to extract valuable data from it. This
problem, which has been extensively discussed in the mobile agent community,
is closely related to the setting discussed here where we assume (one) random
execution errors or errors from unknown sources. In the examples presented in
section 4 we are pursuing an approach where robustness can be obtained through
protocol transformations "in the clear". However, an encoded (or, as in [13], an
"encrypted") execution is also conceivable and would rejoin the approach taken
in quantum computing, as described below.

The detection of incorrect execution also belongs to the context of malicious hosts and our quest for resilient protocols. In [14] a watermark is added to the data which permits to verify whether a remote operation was duly executed. This is structurally similar to the technique used further down in an example where each operation has side effects which serve as hooks for detecting the malfunctioning of an execution circuit.

2.5   Fault Tolerant Quantum Computing

A key aspect of quantum computing is that quantum state (e.g. in a qubit) should not interact with its environment in an uncontrolled way. Unlike classical computers, where state can be measured, restored and copied, it is not possible to copy quantum state or to remove inevitable noise from quantum operations by some threshold scheme. However, it was shown that with an appropriate encoding of a qubit as a codeword over several qubits, it is possible to implement error *detection* capabilities (a) with quantum operations, (b) such that the detection circuitry can also be subject to potential errors and (c) that errors can be *corrected* – again with quantum operations – permitting arbitrarily long quantum computations [15].

This corresponds exactly to what we aim at in this paper: Our goal is to obtain intrinsically robust protocol implementations where errors can occur in the "core" protocol implementation *as well as* in the detection and correction part. However, one difference to the fault tolerant quantum computing approach is that we do not target a logical gate level abstraction on top of which the full (communication) software hierarchy could be stacked. Instead, we wish to expose the unreliability of execution to the highlevel protocol description and handle it inside the protocol. That is, we do not want to (re-)implement basic logic gates and construct computation networks out of them as quantum computing theory does. Another important difference is that we are interested in *software*, and in *arbitrary interconnections between code pieces*, instead of fixed hardware circuits.

2.6   Resilience and Self-healing Properties of Biological Systems

Emergent behavior in biological systems leads to self-regulatory feedback systems that are robust to external perturbation and to failure of its constituent parts [16]. These systems tend to be highly decentralized, the emergent behavior resulting from simple interactions among autonomous agents that make decisions based solely on local views. These simple agents are often anonymous and non-specific, leading to intrinsic fault tolerance and self-healing properties, as other agents can easily take over the roles of failed or missing ones. Several examples can be cited, such as evolutionary selection with survival of the fittest, flock of birds, colonies of social insects, etc. In this section we concentrate on the biochemical processes occurring in cells, which inspired the Fraglets paradigm.

Within a living cell or microorganism there are several chemical processes responsible for maintaining the various cellular functions. These processes can be represented by graphs (networks) where the nodes are the chemical compounds and the links are the reactions that transform these compounds. Biochemical pathways describe the sequence of chemical reactions inside a biochemical network. Among the numerous cellular biochemical networks we can distinguish metabolic networks, responsible for the cell's energy cycle.

These cellular processes are known to be highly robust against mutations, due to several redundancy and diversity mechanisms, such as [17]: the presence of multiple genes with similar functions; interactions among genes with unrelated functions, such as in the case of recessive mutations; the existence of alternative routes in large metabolic networks; the scale-free nature of metabolic networks [18], which are dominated by a few highly connected hubs, while the vast majority of the nodes have a small number of connections, making them inherently robust to random errors.

2.7   Core Wars

Finally we mention "core wars" [19]: Two programs, which share the same random access (core) memory, struggle for survival by attacking the other program through tampering with its instructions and/or by evading attacks through dislocation. Various robustness and self-healing strategies have been proposed for this rather specific context and the associated virtual machine.

3   Parallel and Dynamic Execution Models for Protocols

The traditional implementation of communication protocols with sequential processing of instructions is hard if not impossible to "robustify": An error in a single instruction will most likely disrupt the fragile execution path. Instead, we seek an execution environment where several fine-grained activities can go on in parallel and which can serve as a backup. In this section we present three systems which permit a more parallel expression of communication software.

3.1   Gamma

Back in 1986, Banâtre and Métayer proposed Gamma [20], a programming formalism based on a chemical reaction metaphor. A good overview of the topic with its many ramifications can be found in [21], and a recent update in [22]. Gamma computations consist in "chemical reactions" which consume elements of a multiset data structure, and produce new elements to the multiset. This model enables highly parallel programs to be expressed in a way that is very close to their specification. The authors show that this property makes gamma systems particularly suited as a basis for automated program synthesis.

Many extensions and variations of the basic Gamma system have been proposed, for instance, the Chemical Abstract Machine (CHAM) [23] and Membrane or P systems [24].

These chemical execution models have been applied to diverse fields [21] such as image processing applications, operating systems, compilers, dynamic software reconfiguration [25], multi-agent systems [26] and distributed computing [27]. More recently, γ-calculus has been introduced as a formalism that extends the original Gamma model to a higher-order calculus. In [28] this new calculus has been applied to specify Autonomic Computing systems, including a mailing system as an example. However, to the best of our knowledge, such models have not yet been used to create or reconfigure network protocols.

3.2   Communicating Rule Systems

A formal framework that explicitly addresses communication software is the *Communicating Rule Systems (CRS)* by Mackert and Mackert [29]. Basically a condition/event type of a system, it potentially permits to capture execution variety at the level of single rules such that alternative rules could take over should another rule become unavailable. The rule base, however, is static (due to the author's interest in protocol validation) and becomes a limitation when we want to modify or restore a protocol implementation at run time.

3.3   The Fraglet Paradigm

The Fraglet paradigm [3] has been proposed as part of our search for feasible ways to achieve automated synthesis of protocol implementations. It is based on a chemical model where "molecules" interact with each other or undergo some internal transformation. Formally, it is an instance of Gamma systems [20, 22], described above. Like the higher-order γ-calculus [22, 28], Fraglets explicitly represent code and data in a unified way. The code itself is part of the multiset, a metaphor which is even closer to real chemical systems when compared to the original Gamma model [20, 21]. Adopting this specific chemical model (with Fraglets as the only objects) has the benefit of being able to integrate code deployment into protocols in a natural way. As we will see in Section 4, this will also facilitate the production of instruction-failure resilient code.

A fraglet is a string of symbols $[\, s_1 : s_2 : \ldots : s_n \,]$ which represents data and/or protocol logic. It represents, so to speak, a fragment of a distributed computation. Fraglets may reside inside a node's fraglet store or may be carried in packets, where successive fraglet symbols are analogous to successive header fields in today's regular data packets. Upon arrival, a fraglet packet is injected into the local fraglet store or context.

The fraglet processing engine continuously executes tag matching operations on the fraglets in the store, in order to determine the actions that should be applied to them. Fraglet operations, except for the transmission, have the property that they can be carried out in constant time.

Formally, the store is a multiset: several instances of the same fraglet may be simultaneously present. This is indicated by a suffix counter value as in $[\, data : item \,]k$ (meaning that k copies of fraglet $[\, data : item \,]$ are stored in this context).

The fraglet instruction set currently contains two types of actions: transformation of a single fraglet, and reaction between two fraglets. Table 1 shows some transformation rules defined so far. Table 2 shows the reaction rules.

Table 1. Transformation rules

O p	Input	O utput
dup	$[\, dup : t : u : tail \,]$	$[\, t : u : u : tail \,]$
exch	$[\, exch : t : u : v : tail \,]$	$[\, t : v : u : tail \,]$
new	$[\, new : t : tail \,]$	$[\, t : n_{i+1} : tail \,]$
split	$[\, split : t : \ldots : * : tail \,]$	$[\, t : \ldots \,], [\, tail \,]$
send	$_A[\, send : B : tail \,]$	$_B[\, tail \,]$ (unreliably)

Table 2. Reaction rules

O p	Input	O utput
match (merge)	$[\, match : s : tail_1 \,],$ $[\, s : tail_2 \,]$	$[\, tail_1 : tail_2 \,]$
matchp (persist)	$[\, matchp : s : tail_1 \,],$ $[\, s : tail_2 \,]$	$[\, tail_1 : tail_2 \,]$ $[\, matchP : s : tail_1 \,]$

The semantics of the transformation rules are:

- *dup*: Duplicates the symbol at the third position (u); the second field (t) becomes the new fraglet's head symbol.
- *exch*: Swaps the symbols at the 3rd and 4th position (u and v respectively).
- *new*: Creates a new symbol n_{i+1} which is unique in this context.
- *split*: Breaks the fraglet into two parts at the first marker position ($*$).
- *send*: Sends the fraglet unreliably to the destination context specified by the second symbol (B). The subscript prefix $_N[\ldots]$ denotes the context where the fraglet is stored.

We introduced two simple reaction rules, listed in Table 2. The "merge" instruction ($match$) concatenates two fraglets with matching tags. The "persist" variant ($matchP$) moreover keeps a copy of the initial $[\, matchP : \ldots \,]$ fraglet in the store, thus acts like a catalyst.

It is important to emphasize that our research in defining the instruction set is still in progress, thus the current state should be interpreted as a snapshot of an evolving work. In spite of this apparent limitation, in [3] two protocol implementations using fraglets have been shown: a very simple confirmed delivery protocol and a more complex flow control protocol with send credit and packet reordering.

4   Resilient Protocols

In this section we demonstrate a few simple computation tasks in a communication context whose implementation is robust against partial erasure of their code base. We start with our definition of protocol robustness and instruction failure, and a discussion of appropriate languages to achieve robustness to instruction failure. Then we show two examples of resilient programs at the instruction level: a "signaling frequency doubler" and a confirmed delivery protocol. We then discuss the resulting code and insights gained from this exercise.

4.1   Robustness

Today we have a methodology of designing protocols that makes rather strong assumptions on the reliability of the executing components. The single execution environments are presumed to be stable: Reliable protocols that execute on these components are supposed to handle "only" the unreliable aspects of networking, like broken links, lost packets or transmission errors. That is, robustness applies to the harsh (communication) environment in which computers operate, not to the execution support itself.

Here, we redefine protocol resilience, or protocol robustness, as the ability to survive instruction failures: Even when portions of the protocol's implementation are lost (or duplicated, or changed), the protocol is still able to provide the intended service, albeit with some loss of efficiency. Ideally, a robust protocol implementation will not only be able to detect but also to recover from code losses (self-healing).

Our definition of "protocol robustness" is linked to an implementation, and therefore to the instruction set it is based on. At this point, a question to be raised is which kind of programming languages are suitable to produce instruction-level failure resilient code. Classical procedural languages such as C and Java are very poor candidates, since each single instruction (e.g. assignment, if-then-else, etc.) is linked to many others via cause-effect relationships. Erase a single assignment and the program is likely to crash entirely. Redundancy cannot be directly applied here.

In this paper we use the Fraglet system as our instruction set and consider examples where redundancy can be applied such that any fraglet rule (which is a fraglet by itself) can be removed without changing the outcome of the computation.

An important property that makes fraglets more suitable to instruction-failure resilience is the integration of code and data into the multiset pool. This enables redundancy of code to be expressed in a natural way, without harmful side effects: among a set of redundant rules that match a given input stream, only one of them will be chosen. If at least one of the rules is present, the program can run smoothly in spite of its other sibling rules being knocked out. This will become more clear with the examples of sections 4.2 and 4.3.

4.2   A Robust Message Doubler with Fraglets

The simple task we want to solve is the doubling of a signal stream: for each message x we want two messages z to leave a node or to be available for further processing inside that node. For instance, the messages x could be the ticks from a Geiger counter sensor or any other source which encodes information as a frequency.

Rewriting the signal x (represented as a Fraglet $[x]$) to the new name can be done by the single (non-robust) fraglet rule

[*matchp* : *x* : *z*]

where the $[x]$ will be replaced by $[z]$. The doubling is done by a slightly larger rule, namely

[*matchp* : *x* : *split* : *z* : * : *z*]

which says that fraglet $[x]$ is replaced by a [*split*...] fraglet. This new fraglet will break apart in the next processing step and produce two $[z]$ Fraglets, which is the desired effect.

Unfortunately, this program is not robust: Removing the rule will effectively erase the program. A trivial way to obtain robustness in the Fraglet is to double the rules:

[*matchp* : *x* : *split* : *z* : * : *z*]
[*matchp* : *x* : *split* : *z* : * : *z*]

Here, any one rule instance can be removed, leading to the other rule to be invoked twice as much as before. This seems to satisfy the resilience condition: However, it is not possible to determine, from the program's output, whether a fraglet (rule) has been lost or not. After the first loss, the program is no longer resilient, although it continues to run and produce results, until the single remaining copy is also lost. In order to achieve long-term resilience, the loss of the first Fraglet must be signaled, permitting to trigger a self-healing process.

Therefore we need an implementation such that the side effect of a loss is not harmful to its intended core functionality, but permits to identify which fraglet rule became unavailable, operating as a signal that can be used as input for a self-healing process.

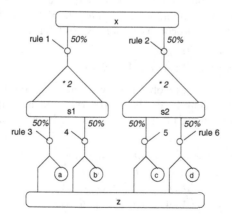

Fig. 1. The processing of fraglets for the robust doubler program

The next version of the doubling program, also shown in Figure 1, is:

Fraglet rule 1: [*matchp* : *x* : *split* : *s*1 : * : *s*1]
2: [*matchp* : *x* : *split* : *s*2 : * : *s*2]
3: [*matchp* : *s*1 : *split* : *z* : * : *a*]
4: [*matchp* : *s*1 : *split* : *z* : * : *b*]
5: [*matchp* : *s*2 : *split* : *z* : * : *c*]
6: [*matchp* : *s*2 : *split* : *z* : * : *d*]

In this case, we have *two different* rules that transform an input fraglet *x* into a *split* fraglet. This means that with a 50% chance, one of these two rules will be picked. Depending on which rule was picked, either two [*s*1 : ...] or two [*s*2 : ...] fraglets will be produced. At this level we have again two rules which transform an [*s*1 : ...] into two different [*split* : ...] fraglets. In fact, it is at this level that the doubling of the original messages is happening: half of the [*x*] fraglets will become [*s*1 : ...] fraglets, but because of the two levels and branches of [*split* : ...] (resulting in a quadrupling), we have an overall doubling of the incoming flux of [*x*] fraglets.

One can observe that removing any one of these 6 fraglet rules will not change the net outcome. If the first rule is removed, the second rule will react with 100% of the influx: for each [*x*] fraglet there will be two [*s*2 : ...] fraglets which each produce a [*z*] fraglet as desired. Similarly, removing any of the other rules will just lead to its "homologue" rule to take over all intermediate fraglets instead of processing only 50% of them.

Note that this implementation generates additional "side effects" in form of the fraglets named [*a*], [*b*] etc. These signals can be used to react on the loss of a rule, as is further discussed in section 4.4.

4.3   A Robust Confirmed Delivery Protocol

This example is a simple confirmed delivery protocol (CDP) already shown in [3]. Node *A* sends an input *data* fraglet to node *B*; when *B* receives the data it delivers it to the application and returns an *ack* fraglet to node *A*. One possible implementation of this protocol, in its non-resilient form, is:

$_A$ [*matchp* : *cdp* : *send* : *B* : *deliver*]
$_B$ [*matchp* : *deliver* : *split* : *send* : *A* : *ack* : *]

The resilient version can be obtained by simply applying the same technique used for the doubler, that is, duplicate the rule and amend it to generate a unique symbol that allows to identify which rule has been executed. The resulting protocol becomes:

$_A$ [*matchp* : *cdp* : *split* : *a* : * : *send* : *B* : *deliver*]
$_A$ [*matchp* : *cdp* : *split* : *b* : * : *send* : *B* : *deliver*]
$_B$ [*matchp* : *deliver* : *split* : *c* : * : *split* : *send* : *A* : *ack* : *]
$_B$ [*matchp* : *deliver* : *split* : *d* : * : *split* : *send* : *A* : *ack* : *]

Actually it is possible to apply the same transformation to any *matchp* rule in order to make it robust. This is a straightforward way to build resilient programs out of normal ones.

4.4   Discussion

The main "trick" of our demo programs is the doubling of rules which compete against each other. This so called "soup" aspect of chemical execution model makes it much easier to write robust code which continues operation despite some losses: any sequential execution model where the control flow must pass through a single instruction is potentially not amendable to robustness, because there is no fallback execution path.

Self-healing. An important aspect of our program is that it produces a stream of signalling symbols, the fraglets a, b, c and d in the examples. These can be used to monitor the health of the program and to trigger repair mechanisms. Adding a few "cleaning" rules will remove these signal streams in case everything works well. For the doubler program, the cleaning rules are:

$[\,matchp : a : match : c\,]$
$[\,matchp : b : match : d\,]$

And for the CDP program, they are:

$_A\,[\,matchp : a : match : b\,]$
$_A\,[\,matchp : b : match : a\,]$
$_B\,[\,matchp : c : match : d\,]$
$_B\,[\,matchp : d : match : c\,]$

As soon as one of the Fraglet rules is removed, this balance is disturbed: By looking at the relative weights of debris ($[a]$, $[\,match : a\,]$ etc.), one can infer for almost all cases which rule was removed. We note here that despite all these detection activities, the doubler program continues to produce twice as many $[z]$ as $[x]$ Fraglets, and CDP continues its normal flow of data and acknowledgments.

The design of control loops – which for example regenerate a lost $[\,matchp : \ldots\,]$ rule – is not trivial and, as our first experiments show, will probably lead to solutions where perfect robustness is not achievable anymore because the result stream can show temporary distortions. Nevertheless, the resulting program performs in an elastic way without fully disrupting the processing. Potentially, one should be able to produce programs which regain correct status after the healing process and given that the error rate is not too high.

However, the exact way to produce healing code able to reconstruct the lost rule is still work in progress. One possibility is to copy one of the remaining rules and rewrite its unique identifier symbol to produce another redundant rule. Another issue is that ideally, the healing code itself should be resilient. Therefore it does not suffice to create a healing "meta-level" that regenerates lost fraglets: a truly self-healing solution should be self-contained. Such a self-contained solution could be obtained in two ways: either by rewriting every program to become resilient and self-healing in itself, or by finding a single "healer" program that is self-healing by itself, i.e. without relying on a meta-level or on special platform features. This program could be used to repair other programs to achieve full self-healing ability, without leading to infinite regression.

Resource Control. In a fully self-healing process we have to deal with resource control issues: unlike the simple demo examples above, which react on the *loss* of fraglet rules, one also has to anticipate an *excess* of rules. Instead of introducing logic to remove superfluous code, we consider the continuous generation of code on the fly and on demand, all new code being automatically consumed after it has been processed. This would lead to a completely dynamic program where the [*matchp* : ...] rules would be replaced by [*match* : ...] rules. These later fraglets would have to be constantly regenerated, thus leading to a program that would be constantly rewriting itself according to its own sensed performance. Metaphorically, we need transcription signals which control the gene expression.

Distribution. As mentioned in the introduction, resilience can be put in a communication context. By changing the intermediate fraglet types such that they undergo a "transmission"-transformation, one can have different parts of the process to occur on different nodes. For the doubler example, assuming that we have four nodes N, M, O and P, we would have the rules:

$_N$ [*matchp* : x : *send* : M : *split* : $s1$: $*$: $s1$]
$_N$ [*matchp* : x : *send* : O : *split* : $s2$: $*$: $s2$]

$_M$ [*matchp* : $s1$: *split* : *send* : P : z : $*$: *send* : P : a]
$_M$ [*matchp* : $s1$: *split* : *send* : P : z : $*$: *send* : P : b]

$_O$ [*matchp* : $s2$: *split* : *send* : P : z : $*$: *send* : P : c]
$_O$ [*matchp* : $s2$: *split* : *send* : P : z : $*$: *send* : P : d]

$_P$ [*matchp* : a : *match* : c]
$_P$ [*matchp* : b : *match* : d]

From node N we rewrite [x] fraglets into fraglets that transfer themselves to nodes M and O and split there. On these nodes, the doubling is performed and the resulting stream is redirected to node P. Note that here we start to blur the distinction between traditional robustness protocols, as the crash of a node M (resulting in removing two rules) would be observable by an imbalance of the monitoring signals and could trigger the necessary repair actions.

5   Conclusions and Outlook

In this paper we have demonstrated simple programs that continue to perform their task despite the removal of any of their instructions. This "intrinsic" robustness is different from the usual extrinsic fault-tolerance approaches as it weaves self-monitoring and self-healing into the proper processing.

The ultimate goal is to create software for autonomic networks which is resilient to accidental or malicious code manipulation and execution problems. In this paper we have shown only the first step towards this goal: instruction-failure resilience, which is achieved using a "chemical" protocol representation and execution model where fallback actions and action equilibria can be easily expressed. Our examples were based on the Fraglet instruction set and multiset memory which has the same fine grained parallelism as the formal Gamma model.

The list of potential research topics in this area is immense. Resilience should be extended to a more general scope beyond single instruction failure: It should be possible to apply the insights from single-instruction knock-out experiments to *networks of components* where medium sized software elements inside a node can crash in a globally recoverable way, or where the amount of elements running concurrently inside the whole network must be controlled. Another important issue is a methodology for transforming existing protocols into a robust, self-healing implementation.

This latter aspect is still work in progress, namely to generate self-healing implementations able to react on the anomaly signals produced by the resilient code, and to regenerate the code base according to these signals. Realistic synthesis methodologies for intrinsically robust protocols will certainly take a long time to mature, even more for self-healing protocols where more insights are needed into "code dynamics".

References

1. Smirnov, M.: Autonomic Communication: Research Agenda for a New Communication Paradigm. White paper, Fraunhofer FOKUS (2003)
2. Zambonelli, F., Gleizes, M.P., Mamei, M., Tolksdorf, R.: Spray Computers: Frontiers of Self-Organization. In: Proceeding of 1st International Conference on Autonomic Computing (ICAC'04), New York, USA (2004) 268–269
3. Tschudin, C.: Fraglets - a Metabolistic Execution Model for Communication Protocols. In: Proceeding of 2nd Annual Symposium on Autonomous Intelligent Networks and Systems (AINS), Menlo Park, USA (2003)
4. Andersen, D.G., Balakrishnan, H., Kaashoek, M.F., Morris, R.: Resilient Overlay Networks. In: Proceedings of the 18th ACM Symposium on Operating Systems Principles (SOSP 2001), Banff, Canada (2001)
5. Jalote, P.: Fault Tolerance in Distributed Systems. ISBN 0133013677. Pearson Education (1994)
6. Dialani, V., Miles, S., Moreau, L., Roure, D.D., Luck, M.: Transparent Fault Tolerance for Web Services based Architectures. In: Proceedings of 8th International Europar Conference (EURO-PAR'02), Paderborn, Germany (2002) 889–898
7. Mullender (Ed.), S.: Distributed Systems, Second Edition. ACM Press, Addison-Wesley (1993)
8. Torres-Pomales, W.: Software Fault Tolerance: A Tutorial. Technical Report TM-2000-210616, NASA (2000)
9. Whisnant, K., Kalbarczyk, Z.T., Iyer, R.K.: A system model for dynamically reconfigurable software. IBM Systems Journal **42** (2003) 45–59

10. Blum, M., Luby, M., Rubinfeld, R.: Self-Testing/Correcting with Applications to Numerical Problems. Journal of Computer and System Sciences **47** (1993) 549–595
11. Wasserman, H., Blum, M.: Software Reliability via Run-Time Result-Checking. Journal of the ACM (JACM) **44** (1997) 826–849
12. Franklin, M., Garay, J., Yung, M.: Self-Testing/Correcting Protocols. In Jayanti, P., ed.: 7th International IS&N Conference on Intelligence in Services and Networks (ISN'00). Springer-Verlag LNCS 1693, Bratislava (1999) 269–283
13. Sander, T., Tschudin, C.: Towards mobile cryptography. In: Proceedings of the IEEE Symposium on Security and Privacy, Oakland, CA, USA, IEEE Computer Society Press (1998)
14. Sander, T., Tschudin, C.F.: On software protection via function hiding. Lecture Notes in Computer Science **1525** (1998) 111–123
15. Preskill, J.: Fault-Tolerant Quantum Computation (1997) arXiv:quant-ph/9712048.
16. Anthony, R.J.: Emergence: A Paradigm for Robust and Scalable Distributed Applications. In: International Conference on Autonomic Computing (ICAC-04), New York, USA (2004)
17. Fontana, W., Wagner, A.:   Mutational Robustness, Modularity and Evolvability.   Research focus area robustness, Santa Fe Institute (2002) http://www.santafe.edu/sfi/research/focus/robustness/projects/ mutationalRobustness.html.
18. Jeong, H., Tombor, B., Albert, R., Oltvai, Z., Barabási, A.L.: The large-scale organization of metabolic networks. Nature **407** (2000) 651–654
19. Dewdney, A.K.: Recreational Mathematics – Core Wars (May 1984) Scientific American. See also http://www.koth.org/.
20. Banâtre, J.P., Métayer, D.L.: A new computational model and its discipline of programming (1986) Technical Report RR0566, INRIA.
21. Banâtre, J.P., Métayer, D.L.: Gamma and the chemical reaction model. Internal publication pi-984, INRIA (1996)
22. Banâtre, J.P., Fradet, P., Radenac, Y.: Principles of chemical programming (2004) Fifth International Workshop on Rule-Based Programming (RULE'04).
23. Berry, G., Boudol, G.: The chemical abstract machine. Research report rr-1133, INRIA Sophia Antipolis (1989)
24. Paun, G.: Computing with Membranes. Journal of Computer and System Sciences **61** (2000) 108–143
25. Wermelinger, M.A.: Specification of Software Architecture Reconfiguration. PhD dissertation, Universidade Nova de Lisboa, Lisbon, Portugal (1999)
26. I. Stamatopoulou, M. Gheorghe, P.K.: Modelling of dynamic configuration of biology-inspired multi-agent systems with communicating X-machines and population P systems. In: Fifth Workshop on Membrane Computing (WMC5), Milan, Italy (2004)
27. Syropoulos, A.: On P systems and distributed computing. In: Fifth Workshop on Membrane Computing (WMC5), Milan, Italy (2004)
28. Banâtre, J.P., Radenac, Y., Fradet, P.: Chemical specification of autonomic systems. In: Proc 13th International Conference on Intelligent and Adaptive Systems and Software Engineering (IASSE'04). (2004) 72–79
29. Mackert, L.F., Neumeier-Mackert, I.B.: Communicating Rule Systems. Protocol Specification, Testing and Verification VII, Proc IFIP WG6.1, 7th International Conference on Protocol Specification, Testing and Verification, Zurich, Switzerland (1987)

Putting Meaning into the Network: Some Semantic Issues for the Design of Autonomic Communications Systems

Simon Dobson

Department of Computer Science,
University College, Dublin IE
simon.dobson@ucd.ie

Abstract. Traditional network abstractions follow a layered model in
which a sub-system interacts with other network components through
very narrow interfaces. We content that this model is weak both in providing clear models of end-to-end properties and allowing adaptation to
the more abstract properties of systems. We propose instead a graph-
centric, contextual abstract model in which sub-systems can relate to
other components at a wide range of semantic levels. We explore the implications such a model would have for network technology, applications
and users, and identify some of the major research challenges it poses.

1   Introduction

Networks are complex creations driven by equally complex software stacks, so
a critical design goal is to minimise complexity for both network and application developers. The traditional approach uses some form of layering, allowing
concerns to be separated behind narrow syntactic and semantic interfaces. This
allows individual layers to be modified, extended and replaced without affecting
the other layers, and crucially allows networks to evolve without dramatically
affecting applications.

Despite its success, however, this trend may be criticised as providing too
narrow an interpretation of the information that can usefully be made use of at
a particular layer of abstraction in a complex software system. By reducing the
information available to a minimum in the interests of simplicity, it is possible
that some opportunities for optimisation are lost. In particular, given the rise of
pervasive computing systems, we would contend that **contextual information
of vital use in adapting the behaviour of a network to its use and
environment is being neglected**, and that this acts as a brake on the creation
of self-managing, self-adaptive autonomic communication systems.

In this paper we make the case for modifying (and perhaps eventually revers-
ing) the trend towards layering, and advocate instead *increasing* the amount of
information available to a network sub-system about the content it is carrying
and the context in which that content will be used. We argue that this view
of the network as an equal partner in interactions – rather than as a simple

M. Smirnov (Ed.): WAC 2004, LNCS 3457, pp. 207–216, 2005.
© IFIP International Federation for Information Processing 2005

packet-carrier – is an appropriate reaction to the desire for autonomic communication systems that facilitates a range of optimisations currently difficult to accomplish in a scalable fashion. We explore the impact that a flatter model has on networks, applications and users, in order both to determine whether such a model has attractions as a research target and, if so, what research challenges it poses.

Section 2 re-visits some of the strengths and weaknesses the current layered approach to network abstractions. Section 3 proposes a graph-based model of information in which network sub-systems exists as sub-graphs rather than layers, which is then analysed in section 4 to determine its impact on some important network-level, application-level and user-level concerns. Section 5 concludes with some suggestions for further exploration.

2   Layered Network Abstractions

Ever since the initial design of TCP/IP there has been a desire among network architects (or at least those involved with internet protocols) to focus on the raw performance of the network. Other architectures that stressed differentiated levels of service (such as X.25) have largely been rejected in favour of the simplicity of the internet model with its single class of packets to be routed efficiently. (For an excellent overview of the history of this process see [1–chapter 5].)

While the single-service packet architecture has proved to be fantastically successful, it is now clear that additional qualities beyond bandwidth are required to support the increasingly wide range of applications for which TCP/IP (and other) networks are being deployed. A good example is the provision of *isochronous* media, in which the temporal properties of delivery are at least as important as the data itself. Indeed, many applications using isochronous media would prefer to drop individual data packets rather than compromise the overall temporal characteristics of the data stream.

Current technical solutions to these issues typically use one of two approaches (or both in conjunction). Firstly, new protocols may be developed that allow the additional characteristics of media to be expressed and supported. The Resource Reservation Protocol (RSVP)[2] is a good example of this. Secondly, networks may be dedicated to particular traffic types such as (for example) video, with the network being dimensioned to ensure delivery. Combination networks are increasingly common, good examples being the MBONE[3] and Voice-Over-IP (VOIP) overlay networks.

The use of protocols and overlay networks is in many ways conditioned by the traditional view of networking embodied in the OSI seven-layer model[4] (figure 1) – physical, data, network, transport, session, presentation, application. (One acronym people sometimes use to remember the layers is "People Design Networks To Send Packets Accurately", which is also a good statement of the traditional view of networks we are criticising!) As with most layered models communications are only allowed between adjacent layers, making it difficult to support "end to end" statements[5]. Although the OSI model's significance is as a conceptual tool rather than a guide to implementing a real network, its notion of layers has become very prevalent.

A key notion in layered architectures is that each layer only "understands" as much about the content it is transporting as its interface allows to be expressed. This is typically very little information expressed at a very low level of abstraction. In this paper, by contrast, we are making the case for leveraging meaning from as wide a range of sources as possible. A key point is that such knowledge is *not* strictly hierarchical: knowledge may potentially have an impact across the spectrum of concerns, and does not follow a strict layered structure.

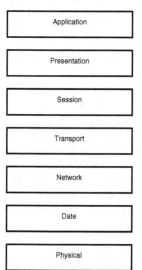

Fig. 1. The OSI reference model

One might argue that layering allows translation of concerns from one domain to another, so layer boundaries provide points at which (for example) session information is translated into transport requests. This means that layer n provides a service abstraction to which higher-level concerns (from layers $n + 1$, $n + 2$ *et cetera*) are mapped. This can result in semantic "squeezing", in the sense that two *different* high-level concerns that happen to map to the *same* layer-n concerns will then be indistinguishable to layers $n - 1$ and so on. This limits the ability of lower layers to react to higher-level concerns.

Conversely, suppose a concern at layer n could be used to inform the adaptation of layer $n - 2$. We may understand this adaptation by saying that layer $n - 2$ responds to certain properties at layer n. However, the concern is mediated by layer $n - 1$, and so must be presented to layer $n - 2$ in layer $n - 1$'s terms. Moreover, for the system to behave predictably this translation must be exact: layer $n - 2$ must adapt *as if* it responded to layer n, although it in fact responds to layer $n - 1$.

In a traditional network, layering is used to *simplify* the way in which concerns interact. In a more complex network, however, layering may *complicate* interactions by introducing translations that are not completely meaning-preserving. One might speculate about a data layer that can account for application preferences in terms of jitter and frame dropping – but which is unable to do so because the intermediate layers do not present these concerns in a coherent form suitable for decision-making.

3   Meaning in the Network

Our concern, then, is that a layered architecture may not provide sufficiently rich access to available information. In traditional networking this is not a problem, and information can be provided implicitly through the development of new protocols *et cetera*. The question we now turn to is whether this remains true as we move towards more self-managing network architectures operating in richer information spaces.

3.1   The Limits of Layered Access to Information

TCP/IP is a general-purpose protocol, specified independently of any particular content. While the same is also true for RSVP, the latter's main purpose is to introduce additional "metadata" into the network – in this case the qualities of service required by isochronous media. At a transport level this takes the form of new message types, new processing rules and so forth; at an application level these manifest themselves in the improved delivery of time-dependent content. Clearly there is a close semantic link between these different levels: one may however question both whether hard-wiring these decisions in protocols is the more appropriate way to proceed, and whether the link can be maintained as the system evolves given the information that is explicitly available for decision-making.

It is important to recognise the distinction between *content* and *use* when discussing networks. The former represents the inherent information being represented and transferred; the latter represents a user task or collection of requirements involving that content. A network protocol is typically more concerned with use than with content: one may transfer the a song with RSVP when it is being played but with HTTP or FTP when it is being archived or mirrored. In the second case, even isochronous media do not place isochrony requirements on the network, and it is desirable to avoid the additional overheads involved. It is not clear that the user or application level should make (or indeed will be able to make) this choice accurately.

In fact there is generally more layering taking place than is apparent in the OSI (or any other) model. Consider, for example, the common practice of securing a data stream using SSL. At one level it is possible to combine SSL with RSVP, by running the SSL packets over the resource-managed connection; at another level, however, this changes the behaviour of the data stream by adding a new layer of processing for encryption and decryption of which RSVP may not be aware and which it may not account for in setting its connection parameters.

Although layering allows individual sub-systems to be changed with minimal disruption, the operative word is *minimal* disruption – which is not the same as *no* disruption. Adapting a system by changing a layer (or component within a layer) may have "knock-on" effects elsewhere in the system. A good example is switching from one encryption system to another, stronger version as content changes, which impacts the transport parameters. Layering can therefore give a false sense of stability in an adaptive system.

More generally, there is a certain arbitrariness in discussions about using networks, protocols and router-based routing decisions in managing networked data streams. It is easy to construct scenarios in which the optimal transport policy – frame size, latency, bandwidth, sensitivity to dropped data, isochrony requirements, security and confidentiality, *et cetera* – depends critically on a wide variety of user-, application- and task-level details. This sensitivity to *context* is a familiar feature of pervasive computing systems: it may be a critical component of next-generation networks too.

3.2   Meaning and Context

Context, in its most general form, may be taken as **the environment in which a system operates understood symbolically**. The concept is used extensively in pervasive computing, where the goal is to make computing devices more adaptive to their users' tasks and requirements as these change over the course in interaction.

A key observation about context is that it is non-hierarchical. One will often encounter links between facts at different conceptual levels that – if captured – illuminate facets of the environment that can prove extremely useful in reasoning and processing. More importantly, context-aware computing makes clear the subtle relationship that exists between users and information. It is not the case that a single piece of information will always be used in the same way; nor is it the case that information has constant relevance to users, or will always be presented at full fidelity, or must always be accessible, and so forth. Pervasive computing is fundamentally concerned with delivering the correct service to the correct user at the correct place and time, and in the correct format for the environment[6]: it is the use of richly interconnected information, and not necessarily constant connectivity or quality of service, that constitutes its chief strength.

Viewing a network as a component of a context-aware system leads immediately to questions of how the network can adapt to the wider context, and how it can be used to inform other components about that context. One may thus see the network – as with most other components of a pervasive computing system – as both a producer and consumer of context.

This leads us to a position contrary to the layered approach of section 2. Instead of viewing a network system as constructed of layers, what happens if we view the system as a *network itself* in which no structure is unduly privileged? This is show schematically in figure 2. The important difference between this view and the layered view is that information can flow directly between different parts of the system – the sub-systems are *sub-graphs* rather than layers.

The significance of this change is two-fold. Firstly, it "flattens" the space of information which a network can access. It reflects a more holistic, and to some extent more trusting, view of information, allowing components to access any information they can usefully process rather than forcing them to use narrow syntactic and semantic interfaces. Secondly, it provides a richly-linked and extensible framework within which to represent *all* the information pertaining to a system's context and configuration in a semantically well-founded manner (for example using RDF[7] to represent assertions about facts). One may then deploy a number of advanced software techniques over this information, the output of which affect the control plane of the underlying network.

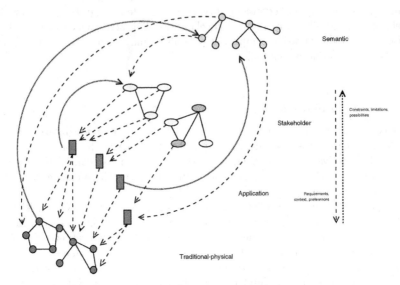

Fig. 2. Network as embedding

As an example, consider the case of an application requirement that a particular data stream be secure against casual observation. The standard solution may be to armour the connection using SSL or similar. However, if a particular network is know to be secure – for example because it is a "hardened" land-line, or a wireless channel that is encrypted with sufficient security automatically at the air interface – then additional armour may be considered unnecessary (and may actually be harmful for security). In a layered model the decision to deploy SSL would almost certainly be made in the top levers, which might not have access to network-layer information about the intrinsic security of a channel; the contextual model potentially allows a decision lower down the stack, accessing the semantic requirements from the higher levels.

The recent interest in the network's "knowledge plane" is a step in the direction we are advocating. The knowledge plane is intended as a single locus for representing and reasoning about higher levels of knowledge in the network, running knowledge-based applications. The important point, however, is not so much the architectural detail but rather to ensure that the management system has access to information at *all* semantic levels within the network – and this might be argued to negate the advantages of plane- or layer-based architectures.

4   Analysis

Conceptual models are important only insofar as they provide insight into design, analysis, teaching or some other mode of understanding. We will therefore explore the model of section 3 above with a view to seeing how a more contextual view of information flow impacts important concerns in autonomic communications, and use these observations to derive technical research questions needing further study.

4.1   Network Concerns

When considering the traditional concerns of networking, the context in which the network is being used may seem too abstract. However, a network is primarily a system for facilitating *human* communication and activities, so the context in which these activities occur is actually of fundamental importance.

What makes the contextual viewpoint different is that it encourages the explicit articulation of information. In the layered world one might deploy RSVP in order to accomplish smooth video display; in the contextual world one might state that (for example) an application is displaying video and requires certain transport characteristics in order to achieve a viable user experience. These are the facts that inform the decision to use RSVP: however, in certain circumstances an alternative decision might equally satisfy the requirements with less overhead, or might be able to meet other requirements that have been implicitly discarded. This looser specification is then an opportunity for adaptation.

While simple data and isochronous media represent two ends of a spectrum of service requirements, there is an important middle ground in which data is transferred using standard protocols and networks but is handled subtly differently depending on its content. Content-based routing (for example [8, 9]) allows routing algorithms to take account of higher-level issues concerning the content and use of the data being routed. A context-based approach can generalise these solutions so that (for example) routing priorities take account of content, user task, information relevance and any other available factors.

If we consider a network as a producer of context, we may integrate its auditing and management functions directly into the overall context model. This is an immediate consequence of the tendency to articulate all available information in a consistent format, and allows management functions to make use of both low- and high-level information on a network's performance. This increases the amount of information available for decision-making, which hopefully leads to improved accuracy.

4.2   Application Concerns

Applications collect together sets of requirements, some of which will impact the network. While traditional packaged applications may not be the best approach to pervasive computing[10], user-visible functionality can still be used to *prima facie* inform the adaptation of the network, and *vice versa*.

A good example of this co-dependence is where an application wants to adapt its presentation according to the bandwidth available. In a layered system this is often problematic, as the application does not have access to the lower layers in which bandwidth concerns are handled. Although some protocols expose the necessary information, many do not. A context system might have the transport sub-system write its current and expected bandwidth availability into the model, where changes can trigger any application expressing an interest in these facts. Conversely a transport system might read applications' expected future transfer requirements from the model and use them to pre-allocate resources.

4.3   Stakeholder Concerns

A major issue in any pervasive system is user acceptance. Any system that is responsive to user context must obviously model that context, and there are few guarantees that the model will be used only for purposes users will accept. The price of adaptation seems to be increased surveillance and more sophisticated models of behaviour.

In many cases it is clear that users will forgo a degree of privacy in exchange for monetary benefit or increased ease of use. It is however equally clear that this trade-off is both difficult to make and problematic to enforce.

There is no obvious answer to these concerns at present. One may observe, however, that systems that articulate changes in parameters that control decisions may also audit how those decisions are made. Extending this paradigm to broader systems may allow a degree of traceability in decision-making that can be used to detect *post facto* some violations of acceptable use.

4.4   Some Questions and Directions

Automated configuration based on requirements sounds very attractive. It does however require a very subtle reasoning process in order to decide on a particular choice of (for example) network protocol from a given constellation of facts, and one might question whether automated reasoning will be adequate to the task – especially given the performance requirements of many network systems. There are also questions about the extensibility of such solutions, given that rule-based systems are often fragile with respect to changes in their rules.

A related question concerns the extensibility of context models. While one may conjecture about networks that use semantic information from applications, this requires as a basis that the network component understands the information model being used by the applications. This seems to imply a considerable degree of coupling between components, which is problematic for both engineering and commercial reasons.

A more semantic issue is the stability of systems reacting to rich models. One can easily conceive of a system that is finely balanced between two possible choices and oscillates between them, incurring costs at each oscillation. This is especially hard to find when the changes are effected by different components reacting in conflicting ways to separate parts of the model.

We highlighted the ability to integrate low- and high-level information into management tools. The benefit of this approach is that is simplifies the expression of end-to-end properties, in that there is a "trace" from high-level requirements to management decisions. Creating this pathway implies that we can provide suitable decision procedures and handle conflicts between requirements and tasks.

The use of pervasive computing as a surveillance tool is a widespread concern, and one which has perhaps not received sufficient attention from the technical communities so far. It goes significantly beyond the normal issues of cryptography, into the realms of traffic analysis used by the military to gain information from who is talking to whom (regardless of whether the actual communications can be read). Users (and corporations) will not use systems that may leak information to marketeers (or competitors, or governments). It is an open question where autonomic communications stands in the space of adaptivity *versus* intrusiveness.

5   Conclusion

We have investigated how the traditional notion of layering in a network may not be optimal for accessing the richer spaces of information that may be available to next-generation networks, and especially in pervasive computing systems. We have argued for a flatter, more extensible context model to allow sub-systems to access information from any appropriate semantic level, and have explored some of the issues that such a model raises.

Our intention in this paper has purely been to explore the attractions and feasibility of applying a more holistic and contextual approach to information to autonomic communications and self-adaptive networks. Our conclusions would be (firstly) that such an approach does have significant attractions, and (secondly) that there are significant challenges in terms of user control, information leakage and the extensibility of decision procedures. Nevertheless we conclude overall that a semantically well-founded approach to autonomic communications based on techniques from contextual modelling is worthy of further attention in the future.

The challenges that we have identified relate primarily to issues of knowledge representation and automated reasoning – what might broadly be called artificial intelligence. Not all the capabilities we have identified exist currently in a form suitable for use in the way we have suggested. However, by developing and applying these techniques to autonomic communication we hope to improve their capabilities for adaptation and management while retaining a degree of confidence in the correctness and compositionality of these capabilities. Both are vital if communications are to become truly autonomic.

Acknowledgements

The work in this paper expands on discussions that took place at a European Union workshop on Autonomic Communications held in Brussels in March 2004, for which the author acted as one of the rapporteurs. Many of the participants raised points that contributed greatly to the ideas presented here, although any misunderstanding of these points is of course due to the author alone.

References

1. Abate, J.: Inventing the Internet. MIT Press (1999)
2. Braden, R., Zhang, L., Berson, S., Herzog, S., Jamin, S.: Resource ReSerVation Protocol (RSVP) – version 1 functional specification. RFC 2205 (1997)
3. Savetz, K., Randall, N., Lepage, Y.: MBONE: Multicasting tomorrow's internet (1995)
4. : Open systems interconnection – basic reference model: the basic model. Technical Report ISO/IEC 7498-1:1994, ISO (1994)
5. Saltzer, J., Reed, D., Clark, D.: End-to-end arguments in system design. ACM Transactions on Computer Systems (1984) 288–288
6. Weiser, M.: The computer for the 21st century. Scientific American (1991)
7. Lassila, O., Swick, R.: Resource Description Framework model and syntax specification. Technical report, World Wide Web Consortium (1999)
8. Debusmann, M., Geihs, K.: Towards dependable web service. In: Proceedings of the 10th IEEE/IFIP International Pacific Rim Symposium on Dependable Computing, IEEE Press (2004) 5–14
9. Yang, C.S., Luo, M.Y.: Efficient support for content-based routing in web server clusters. In: 2nd USENIX Symposium on Internet Technologies & Systems. (1999)
10. Dobson, S.: Applications considered harmful for ambient systems. In: Proceedings of the International Symposium on Information and Communications Technologies, ACM Press (2003) 171–176

Dynamic and Contextualised Behavioural Knowledge in Autonomic Communications

Roy Sterritt[1], Maurice Mulvenna[1], and Agnieszka Lawrynowicz[2]

[1] School of Computing and Mathematics, University of Ulster at Jordanstown,
Newtownabbey, County Antrim, BT37 0QB Northern Ireland
{r.sterritt, md.mulvenna}@ulster.ac.uk
[2] Institute of Computing Science, Poznan University of Technology,
ul. Piotrowo 3a, 60-965 Poznan, Poland
agnieszka.lawrynowicz@cs.put.poznan.pl

Abstract. The conceptual architecture of autonomic communications requires a knowledge layer to facilitate effective, transparent and high level self-management capabilities. This pervasive knowledge plane can utilise the behaviour of autonomic communication regimes to monitor and intervene at many differing levels of network granularity. This paper discusses autonomic computing and autonomic communication, before outlining the role of behavioural knowledge in autonomic networks. Some research issues, in particular the concept of dynamic context as a method to acquire knowledge dynamically that will help to facilitate a successful realisation of the knowledge plane are explored and discussed.

1 Introduction

An EU FET brainstorming workshop in July 2003 to discuss novel communication paradigms for 2020 identified '*Autonomic Communications*' as an important area for future research and development [1]. This can be interpreted as further work on self-organizing networks, but is undoubtedly a reflection of the growing influence of IBM's Autonomic Computing initiative launched in 2001 [2]. In effect, autonomic communications has the same motivators as the autonomic computing concept with particular focus on the communications research and development community. Goals highlighted at this initial workshop were to understand how an autonomic network element's behaviours are learned, influenced or changed, and how in turn, these effect other elements, groups and networks. The ability to adapt the behaviour of the elements was considered particularly important in relation to drastic changes in the environment such as technical developments or new economic models [1].

At the heart of autonomic communications are *selfware* principles and technologies that will create the autonomic network. They borrow largely from autonomous distributed systems research and non-conventional networking (ad hoc, sensor, peer-to-peer, group communications, active networks and so on), among others [3]. In addition to this a new construct, a knowledge plane, has been identified as being required to act as a pervasive system within the network that builds and maintains high level models of what the network is supposed to do in order to provide the

M. Smirnov (Ed.): WAC 2004, LNCS 3457, pp. 217–228, 2005.

communications services and advice to other elements in the network [4]. It is generally considered that this knowledge plane will rely on the tools of AI and cognitive systems (to meet the uncertainties and complexities of this goal) rather than traditional algorithmic approaches [4][5].

This paper motivates the proposition that the successful creation of autonomic communications, and in particular the knowledge plane, requires the ability to possess context awareness and behavioural knowledge from an ethnomethodological perspective. Ethnomethodology is an in-depth study of individuals and groups, their practice, and their artefacts in the context of their normal working environment. From this perspective, context is more than just the sum or function of the metrics that are monitored or probed in the environment.

The paper outlines the area of autonomic computing and autonomic communications before beginning to discuss the role of knowledge in autonomic communications. The remainder of this vision paper is a discussion on how knowledge may be used within the knowledge plane of autonomic communications. In particular, we examine the mechanism of dynamic context as a framework for the generation, use and execution of knowledge in autonomic networks.

2 Autonomic Computing

The autonomic metaphor, based on the human body's autonomic or self-regulating and protection system, strives to achieve systems which will maintain themselves through the use of an autonomic element consisting of an autonomic manager and the managed component. There is a strong requirement for dependability, from single mobile devices running multiple processes through distributed grid applications [6].

The general properties of an autonomic (self-managing) system can be summarised by four objectives; self-configuring, self-healing, self-optimising and self-protecting and four attributes; self-awareness, environment-awareness, self-monitoring and self-adjusting [6]. Essentially, the objectives represent broad system requirements while the attributes identify basic implementation mechanisms. (Since the 2001 launch of autonomic computing the *self-** list has grown substantially yet this initial set still represents the general goal.)

Self-configuring is a system's ability to readjust itself automatically, this may simply be in support of to changing circumstances or to assist in self-healing, self-optimisation or self-protection. Self-healing, a reactive mechanism is concerned with ensuring effective recovery when a fault occurs; identifying the fault and then where possible repair it. Self-optimisation means that a system is aware of its ideal performance, can measure its current performance against that ideal and has policies for attempting improvements. A self-protecting system will defend itself from accidental or malicious external attack. This means being aware of potential threats and having ways of handling those threats. This may include self-healing actions if an attack is successful, and a mix of self-configuration and self-optimisation to increase protection. Finally, these self-mechanisms should ensure there is minimal disruption to users, avoiding significant delays in processing. To achieve these objectives a system must be aware of its internal state (self-aware) and current external operating conditions (environment-aware). Changing circumstances are detected through self-monitoring and adaptations are made accordingly (self-adjusting). As such, a system must have knowledge of its available resources, its components, their desired performance characteristics, their current status, and the status of inter-connections with other systems, along with rules and policies of how these may be adjusted. The ability to operate in a heterogeneous environment requires the use of open standards to understand and communicate with other systems [1].

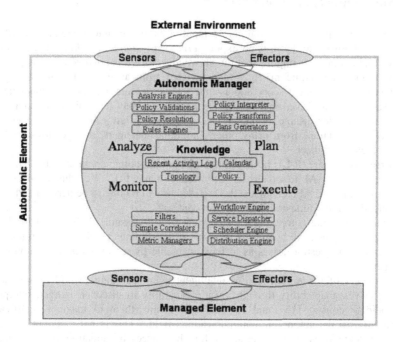

Fig. 1. IBM's view of the Architecture and Components of an Autonomic Element (adapted from [7])

At the heart of any autonomic system architecture are *sensors* and *effectors* [8]. A control loop is created by monitoring behaviour through sensors, comparing this with expectations (historical and current data, rules and beliefs), planning what action is necessary (if any) and then executing that action through effectors [9]. The control loop, a success of manufacturing science for many years, provides the basic backbone structure for each system component [7].

Figure 1 is IBM's view of the necessary components within an autonomic manager. (For an alternative artefacts view, see [10].) It is assumed that an autonomic manager is responsible for a managed element within a self-contained autonomic element. Interaction will occur with remote autonomic managers through virtual, peer-to-peer, client-server [11] or grid [12] configurations.

The monitor and analyse parts of the structure process information from the sensors to provide both self-awareness and an awareness of the external environment. The plan and execute parts decide on the necessary self-management behaviour that will be executed through the effectors. The *MAPE* (Monitor, Analyse, Plan and Execute) components use the correlations, rules, beliefs, expectations, histories and other information known to the autonomic element, or available to it through the knowledge repository within the autonomic manager (AM).

3 Autonomic Communication

The explicit perspective for autonomic computing is that an autonomic element (AE) solely *uses* knowledge; there is no explicit *creator* or *adaptors* of knowledge within the AE architecture. It implies that the knowledge within is *engineered in* as part of the developed autonomic manager (and updated from an external source). If you consider the management scope and assume this autonomic manager's component is for instance a disk drive, engineering the knowledge may be achievable yet if the scope is larger, for instance a higher level manager within a server farm receiving event communications from many other autonomic managers, the scenarios will be too complex to engineer.

AI may assist here. It has been highlighted that autonomic communications has an intrinsic need for AI to create the knowledge plane [4]. Yet this need is not necessarily standard AI techniques; the knowledge will require to be derived and used dynamically, in real-time and in the correct context.

In this autonomic computing view, even if you do assume that AI and machine learning techniques have been used to assist in developing/engineering the rules and beliefs, another question arises as to how adaptable these are within the autonomic manager.

Proponents of the mobile and/or intelligent agent paradigm would present that context drives adaptability through agent's capability to discover, extract, interpret and validate context [13], and as such will enable them to make a significant contribution to the autonomic communications field. This is not in doubt; a wide range of techniques will be required for the successful creation of autonomic communications. Yet will emergent behaviour from autonomic elements agents provide the scope envisaged at the knowledge plane level?

An interesting paper in [14] discusses affect and machine design [15]. Essentially it supports those psychologists and AI researchers that hold the view that affect is essential for intelligent behaviour. It proposes three levels for the design of systems:

1. Reaction – lowest level where no learning occurs but immediate response to state information coming from sensory systems.
2. Routine – middle level where largely routine evaluation and planning behaviours take place. It receives input from sensors as well as from the reaction level and reflection level. This level of assessment results in three dimensions of affect and emotion values: positive affect, negative affect and (energetic) arousal.
3. Reflection – top level receives no sensory input or has no motor output, it receives input from below. Reflection is considered a meta-process, where the mind deliberates about itself. Essentially operations at this level look at the systems representations of its experiences, its current behaviour, its current environment etc.

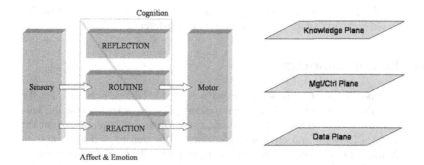

Fig. 2. Intelligent Machine Design three tiers compared with three planes in Autonomic Communications

The affect and emotion debate is not an issue here, it is the three levels that are of specific interest. Although not described in such terms this approach to intelligent design is similar to the proposed scoping of the planes within autonomic communications (Figure 2). Essentially the reaction level may be considered to sit within the data plane and the autonomic network, where for instance under fault conditions automated switching and fail-over may take place, and so on in, monitoring current state of both the network element and its environment with rapid reaction to changing circumstances. The routine level may be considered the management plane, where planning takes place and under fault conditions root cause analysis is performed on the event messages from the data plane. The reflection level may be considered not to reside yet within networks but is akin to the perceived knowledge plane, where it considers the behaviour of the networks and learns new strategies, reflects upon the success of existing strategies and adapts if necessary.

This approach highlights the need for reflection and cognitive strategies to be designed into systems to provide the self-adaptability autonomic property.

Self-adapting behaviour has been classified into three levels by the Smart Adaptive Systems community. These are [16]:

1. Adaptation to a changing environment,
2. Adaptation to a similar setting without explicitly being ported to it,
3. Adaptation to a new/unknown application.

Its seems a difficult task for an autonomic element and indeed the autonomic network and autonomic communications to even conform to level 1 through engineering rules into the autonomic manager. To be classified level 2 is certainty going to entail AI and cognitive approaches.

This section has briefly focused on the initial general designs of an autonomic element to emerge from autonomic computing, key criteria consisting of self-managing (self-CHOP[1] or self-*), AE=AM+ME, MAPE, control loop, sensors+effectors, all reliant on knowledge repository to provide self and environment

---

[1] Objectives: self-Configuring, self-Healing, self-Optimising and self-Protecting.

awareness. These *users* of the knowledge component have yet to identify how this knowledge will be learnt, adapted or even used within different contexts.

4 The Role of Knowledge

Behavioural knowledge and knowledge execution is a vital research area for the successful fulfilment of Autonomic Communication [17]. In order to drive the self-managing capabilities of autonomic communications, there is a requirement for the network to be self-aware and environment-aware. Research on self-awareness in next-generation networks can be driven by attempting to understand the behaviour of the network. To achieve this, the network must have access to various data and knowledge components, on which it can execute and modify its parameters.

The data and knowledge sources are [18]:

- Deriving and using first- and second-order data from the data plane of the network;
- Deriving and using network management data and knowledge from the control plane of the network;
- Deriving and using data and knowledge that comprises the knowledge plane [4] of a network.

The first two data and knowledge sources can be and are employed to varying degrees in network research today. The third area represents a significant advance in research thinking, in that it is primarily inferential and mined knowledge that is discovered by predictive analytic techniques residing on the knowledge plane. These techniques include collaborative filtering [19][20], Bayesian networks [21], clustering [22], classification [23], association rules [24], sequence analysis [25] and content filtering [26] as well as runtime techniques from click stream analysis [27][28].

The knowledge plane must have the capacity to retain and maintain a *network memory*, comprising the data and knowledge sources indicated above. An excellent starting point for this memory will be machine-understandable XML-based syntax, comprising different standards that maintain high semantic integrity and coherence for the knowledge models; for example, the Predictive Modelling Mark-up Language (PMML) [29].

PMML is an XML-based standard developed by the Data Mining Group with the aim of aiding model exchange between different model producers and between model producers and consumers. Most data mining vendors have their own proprietary representations for knowledge discovered using their algorithms. PMML provides the first standard representation that is adhered to by all the major data mining vendors. Being XML-based, models represented in PMML can easily be parsed, manipulated and used by automated tools. The anticipated use of flexible, semantically-rich representational schemes such as PMML within autonomic elements is as a memory for policies and events that provides fast, interactional response in autonomic network environments.

This network memory will be maintained as a discrete ontological construct in the knowledge plane, necessitating new research in network ontologies. This memory is, in essence, a collection of rule sets and mining model result sets that can maintain network policies as well as behavioural descriptions and policies. As such, it is a memory that provides context for measurement. Therefore, via introspection and mediation, the memory can self-adapt to improve performance depending on the context and needs of use.

In order to execute and interact with the network memory, a scalable high-performance engine is required. This is similar in construct to a recommender engine [27][28], in that it is constantly updating the network memory rule bases upon which the application of predictive algorithms on network behavioural data is based.

A key component of this engine is the detection of network trends and subtle changes in data flows. Key research currently under way in concept drift may be the basis for drift detection in autonomic network architectures [30].

There are key challenges in this research sub-area of autonomic communications, including the real time handling and assessment of ensembles of behavioural knowledge to improve network provision and the ability to introspectively measure the performance, accuracy and appropriateness of network performance.

5 Contextualised Knowledge

The autonomic communications knowledge plane not only requires the ability to *use* knowledge but also the ability to *create* and *adapt* it when necessary. A vital aspect to these abilities is to understand the *context* within which that knowledge is framed.

The understanding of *context* has been a significant research area in many fields of computing, in particular AI and ubiquitous computing, for some time now. The term context-aware computing was first introduced by in 1994 [31] as a system's ability to adapt to its location of use and objects (people, devices) in the neighbourhood. It was defined in the context of the systems in which the user employs many different mobile, embedded and stationary computers in different situations and locations over the course of the day. This has evolved within several research fields sharing many common views, including ubiquitous computing [32], pervasive computing [33], ambient intelligence [34], planetary/utility/grid computing and so on.

Many definitions of context-awareness and models of context-aware systems have been proposed, the most popular over-arching perspective that researchers from pervasive computing society employ is to see context as some function or mode of the parameters of the environment, such as time, place, etc. Values of the parameters are acquired by using the predefined set of sensors and then extract features from these low-level sensor readings [35].

Acquiring context is not a straightforward task due to its dynamic nature and the heterogeneous state of data sources. Context can be extracted from low-level sensors and high level managers as well as derived from applications to-date utilising the network. It has been highlighted that the majority of context-aware applications use the data from the sensors later offline through data pre-processing and features extraction [35].

There is no consensus for context representation (capturing, representing and modelling context). Problems concern the fact that acquired information can be strongly heterogeneous and often incorrect, inconsistent or incomplete. A second issue is that it is used in systems in various ways. A substantial amount of different approaches have been proposed to model such contextual information.

Dourish [36] has suggested that the representational approach of context applied by most of the researchers until now interprets the role of context in a different manner than it plays in our everyday life. He proposes instead a new perspective on context-aware computing where the context is perceived much like in social sciences that study the practises of individuals in their normal environment. In his article he examines the problem of context from a high-level, philosophical point of view, enumerates the various philosophical viewpoints, and highlights an approach, which views context as an interactional problem rather than a representational one.

This *dynamic context*, as we term it, contrasts with a majority of the literature concerning context-awareness, particularly engineering approaches that inherit from positivist theoretical tradition which seek objective answers, independent of the detail of particular occasion descriptions of social phenomena. It is a positivist point of view in which we look at things as something to be modelled and encoded. From this perspective, context is a stable feature of the world that is independent of the actions of individuals.

On the other hand, dynamic context proposes to look at the problem of context-aware computing from another, *phenomenological* point of view. In this view, social facts are not pre-given or absolute but are continually negotiated and reinterpreted as a result of interactions hence perception of the world depends on the interpretation of particular individuals.

In dynamic context, it is the activity that generates and sustains the context. So, context arises from the activity, and is actively produced and maintained in the course of the activity. This provides a framework for a method to determine context from activity via behaviour (and measures of behaviour). This framework is a justifiable research goal in autonomic communications.

Although the assumptions enumerated above seem to be a correct way to view the role that context should play in context-aware systems, there are many significant issues concerning how to turn this approach into reality. Dynamic context is only a conceptual view of what context is and formal design guidelines for systems are not presented. It is an interactional model of context, in which the central problem is "how and why, in the course of their interactions, do people achieve and maintain a mutual understanding of the context for their actions?" It can be argued that the difficulty and practical problems of designing context-aware systems has encouraged the pervasiveness of the representational view of context.

Context is hard to recognise and hard to encode. The approach of dynamic context makes this task even harder, because instead of the readings from the set of predefined sensors we have to deal with the features that can be contextually relevant to the particular activity. It cannot be determined *a priori*, before an activity happens. Some features become meaningful for particular sorts of actions - that's why the context should be continually redefined, as such the scope of contextual features should be defined dynamically. That forces the representation to be flexible enough to maintain the changing importance of the features in different types of activities and their dependence together with the possibility to add or delete features. Dourish gives a conceptual idea of how context should be understood and suggests to move the stress from designing how to use the predefined context within a system, but rather how the system can support the process by which "context is continually manifest, defined, negotiated, and shared".

6 A Framework for Using Context

In dynamic context, the activity generates and sustains the context. This fresh perspective is drawn ultimately from social science techniques such as ethnomethodology and ethnography, and explores their usefulness in the increasing number of computer-mediated pervasive and ubiquitous environments. Ethnomethodology simply means the study of the ways in which people make sense of their social world [37]. Ethnography is the in-depth study of individuals and groups, their practices, and their artefacts in the context of their normal work environment [38].

The usefulness of ethnography seems to be that it takes nothing for granted, and the application in anthropology ensures that all details are available for analysis. Ethnography is a contender for a framework in which we seek to discover context from activity. Having the measures of behaviour, we can then try to discover from them the activity and from the activity we can try to discover the context. This leads to the exciting unexplored possibility for a new general framework for context-aware computing.

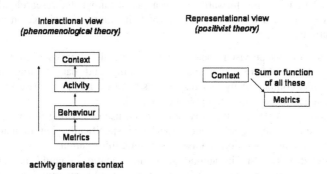

Fig. 3. Interactional Vs Representational view of context-awareness

Having a set of sensors within the autonomic elements, each sensor provides the measures of network, systems and user application behaviour. On the basis of the AE's measurements we can try to discover the application's current activity. Frequent patterns found in the measurements can then be labelled and represent typical activities. Having discovered activity we can try to compute the context generated by the given activity, understand a more precise description of the activity in the form of its goal, and the conditions in which it is executed (that means context viewed by the prism of activity). The general Behaviour-Activity-Context (BAC) Framework for context-awareness is shown in on the left-hand side of Figure 3.

7 Discussion and Proposed Future Emerging Research Agenda

The knowledge plane is a proposed third abstraction in the emerging research area of autonomic communications, adding to the existing data and control/management planes. In their vision paper, the proponents of the knowledge plane discuss broadly how machine learning algorithms can be applied to garner knowledge and increase the self-awareness of the network. How the knowledge plane will be achieved is an open research area, but the remaining discussion examines what role contextualised knowledge may play in autonomic communications.

The paper first focused on a brief review of the general autonomic element designs emerging from autonomic computing noting that the general architecture of the autonomic computing autonomic element would imply it is only a *user* of knowledge with no explicit components for *creating* nor *adapting* knowledge. Agents, AI and cognitive techniques may assist here. It was highlighted that the three tiers; reflex, routine and reflection of the proposed autonomic intelligent machine design have scope commonalities with the data, management and knowledge planes within autonomic communications.

The second EU FET consultation meeting in March 2004 on the subject of autonomic communications [39][40][41] highlighted that self-awareness in autonomic communications must be driven by self-knowledge, specifically by behavioural knowledge. This key area was entitled behavioural knowledge and knowledge execution. The authors put forward several areas of knowledge research which they feel should be pursued to support the use of behavioural knowledge in autonomic communications:

- The use of unsupervised, incremental learning algorithms should be explored. Although there are many machine learning and data mining algorithms available, comparatively few researchers have explored this area, in particular from a pervasive computing perspective [36].
- The second of these is the development and use of existing research and tools that facilitate high-performance operation; specifically, ontological tools to support the incorporation and use of semantic information.
- Knowledge systems in autonomic communications should be capable of practicing introspection. That is, they should measure the degree of correctness of their 'advice' within an autonomic element.
- The knowledge system should be capable of discriminating between conflicting types of advice and selecting or blending advice. This can be explored initially as simple conflict resolution, but a key goal would be the development of managers of ensemble advisors or recommenders within autonomic communications.

More generally, this paper has introduced the concept of dynamic context and advanced the proposition that the successful creation of autonomic communications and the knowledge plane will not only require AI and cognitive approaches but will also require a fuller interpretation of context; in some ways akin to ethnography, building towards the formulation of a novel context-awareness framework; Behaviour-Activity-Context (BAC).

A treatment of the area of context-awareness highlighted the two schools of thought; *interactional* versus *representational*, or *phenomenological* versus *positivist* perspective. This paper supports the interactional view where context is generated and sustained by the activity. This ethnographical-inspired view of the world, which we have labelled dynamic context, should provide the most dynamic knowledge approach for autonomic communications.

This paper explicitly focused on one of the grounding principles to achieve autonomic communications – a new communication paradigm to assist the design of the Next Generation Networks (NGN) – that of contextualised knowledge.

We have proposed a new dynamic context model, based upon on-the-fly, dynamic and lightweight analysis of data in the network, as well as a workable framework for experiments. We propose a research plan that tests the hypothesis that contextualised knowledge can improve the capabilities of a knowledge plane in autonomic communications. The details of the roadmap for this plan will be explored in a future paper.

References

1. EU IST FET, "New Communication Paradigms for 2020", brain storming meeting July 2003, Brussels, Belgium, (report published Sept 2003)
2. P. Horn, "Autonomic computing: IBM perspective on the state of information technology", AGENDA'01, October 2001
3. M Smirnov, R Popescu-Zeletin, "Autonomic Communication", presentation EU IST FET brainstorming meeting Communication Paradigms for 2020, Brussels, July 2003
4. D Clark, C Partridge, JC Ramming, JT Wroclawski, "A Knowledge Plane for the Internet", Proc. Applications, technologies, architectures, and protocols for computer communication, Karlsruhe, ACM SIGCOMM 2003
5. JM Agosta, S Crosby, "Network integrity by inference in distributed systems", NIPS Workshop on Robust Communication Dynamics in Complex Networks, 2003
6. R Sterritt, DW Bustard, "Autonomic Computing - a Means of Achieving Dependability?", Proc. IEEE Int. Conf. Engineering of Computer Based Systems (ECBS'03), Huntsville, AL, USA, , pp 247-25, April 7-11 2003
7. A Ganek, "Autonomic Computing: Implementing the Vision", Keynote presentation at the Autonomic Computing Workshop, (AMS 2003), Seattle, WA, 25th June 2003.
8. AG Ganek TA Corbi, The dawning of the autonomic computing era, IBM Sys J 42(1) 5-18 2003
9. Autonomic Computing Concepts, White Paper, IBM, 2001
10. R. Sterritt, DW Bustard, "Towards an Autonomic Computing Environment", 1st Int. Workshop Autonomic Computing Systems at DEXA'2003 Prague, 694-698 2003.
11. DF Bantz, C Bisdikian, D Challener, JP Karidis, S Mastrianni, A Mohindra, DG Shea, M Vanover, Autonomic personal computing, IBM Sys J 42(1) 165-176 2003
12. G. Deen, T. Lehman, J. Kaufman, The Almaden OptimalGrid Project, IEEE Proc. Autonomic Computing Workshop, (5th AMS), Seattle, WA, pp 14-21, June 2003.
13. A. Zaslavsky, Mobile Agents: Can they assist with Context Awareness?, Proc. 2004 IEEE Int. Conf. Mobile Data Management, 2004
14. IBM Systems Journal, Special issue on Autonomic Computing, Vol. 42, No.1, 2003
15. DA Norman, A Ortony, DM Russell, "Affect and machine design: Lessons for the development of autonomous machines", IBM Sys J, 42(1), 38-44, 2003.

16. D Anguita, "Smart Adaptive Systems: State of the Art and Future Direction of Research", EUNITE, Dec. 2001
17. M Smirnov, 'Area: Autonomic Communications', EU IST FET New Communication Paradigms for 2020 Consultation meeting, Brussels, Belgium. (ver. 02), March 2004.
18. MD Mulvenna, Annex: Comments and background material to topic 8 Behaviour knowledge and knowledge execution in Autonomic Communications, IST FET New Communication Paradigms for 2020 Consultation meeting, Brussels, March 2004.
19. U Shardanand, P Maes, Social information filtering: Algorithms for automating "word of mouth", Proc CHI'95 Human Factors in Computing Systems, 210-217 1995
20. P Resnick, N Iacovou, M Suchak, P Bergstrom, J Riedl "GroupLens: An open architecture for collaborative filtering of netnews", Proc ACM 1994 Conf Computer Supported Cooperative Work, Chapel Hill, NC: ACM, 175-186 1994
21. D Heckerman, D Geiger, D Chickering Learning Bayesian networks The Combination of Knowledge and Statistical Data, Machine Learning, 20 197-243 1995
22. AK Jain, RC Dubes, Algorithms for Clustering Data. Pren Hall, NJ 1998
23. D Mitchie, DJ Spiegelhalter, CC Taylor, Machine Learning, Neural and Statistical Classification, 1994. www.amsta.leeds.ac.uk/~charles/statlog/
24. R Agrawal R Srikant, Fast Algorithms for Mining Association Rules. In Proc. of the 20th Int'l Conference on Very Large Databases, Santiago, Chile, September 1994.
25. AG Büchner, M Baumgarten, SS Anand, MD Mulvenna, JG Hughes, Navigation Pattern Discovery from Internet Data, Advances in Web Usage Analysis and User Profiling, LNCS, Springer-Verlag, 2000.
26. M Mobasher, R Cooley, J Srivastava, Automatic Personalization Based On Web Usage Mining, Communication of ACM, August, 2000, Volume 43, Issue 8
27. MD Mulvenna, SS Anand, AG Büchner, (eds.), Personalization on the Net using Web Mining, Comm. ACM Special Section, 43(8), 122–125, Aug. 2000
28. AG Büchner, MD Mulvenna, Discovering Marketing Intelligence Through Online Analytical Web Usage Mining, ACM SIGMOD Record, 27(4), 54-61, 1998
29. PMML, www.dmg.org
30. M Black, R Hickey, Learning classification rules for telecom customer call data under concept drift. Soft Comput. 8(2): 102-108 (2003)
31. B Schilit, N Adams, R Want, R. Context-Aware Computing Applications. Proc. IEEE Workshop on Mobile Computing Systems and Applications, 1994.
32. M Weiser, The Computer for the 21st Century. Sci American, 265(3), 94-104 1991.
33. W Ark, T Selker, A Look at Human Interaction with Pervasive Computers, IBM Sys J, 38(4), 504-507 1999
34. E Aarts, R Collier, E van Loenen, Bd Ruyter, (Eds.), Ambient Intelligence, 1st European Symposium, EUSAI 2003, Veldhoven, The Netherlands, LNCS 2875 2003
35. R. Mayrhofer, H. Radi, et al. Recognizing and predicting context by learning from user behaviour, Austrian Computer Society (OCG), 2003.
36. P. Dourish, What we talk about when we talk about context. Personal and Ubiquitous Computing 8(1) 19-30, 2004
37. Garfinkel, H., *Studies in Ethnomethodology*, Prentice-Hall, 1967
38. Bowling, A. (1997). *Measuring health: A review of quality of life measurement scales* (2nd ed.). Philadelphia: Open University Press.
39. M Smirnov, Managing Internet complexity in Autonomic Communication, presentation EU IST FET consultation meeting Communication Paradigms for 2020, Brussels, March 2004
40. EU IST FET, "New Communication Paradigms for 2020", Consultation meeting 3-4th March 2004, Brussels, Belgium.
41. F Sestini, 'Situated and Autonomic Communications', EU IST FET New Communication Paradigms for 2020 Consultation meeting, Brussels, March 2004.

Towards Adaptable Ad Hoc Networks: The Routing Experience[*]

Cesar A. Santivanez[1] and Ioannis Stavrakakis[2]

[1] Internetwork Research Department,
BBN Technology, Cambridge, MA 02138, USA
`csantiva@bbn.com`
[2] Department of Informatics and Telecommunications,
University of Athens, 15784 Athens, Greece
`ioannis@di.uoa.gr`

Abstract. Network users not only demand new and versatile application support by the networks but they themselves are becoming part of the network (network routers, caches, processors, etc) by contributing their resources to it and being engaged in ad hoc networking structures. As the large and diverse user population becomes more and more part of the networking infrastructure it is clear that networks will be dominated by a new type of network nodes which are much more nomadic, diverse and autonomic than in traditional networks, creating a fairly diverse – in size and characteristics – networking environment. For instance, low cost/high availability/convenience of wireless devices are expected to lead to the deployment of a plethora of wireless networks for diverse applications: from rescue missions to military communications, from collaborative computing and sensor networks to web browsing and e-mail exchange to real time voice and video communications. Each with different constraints and requirements. And, for each type of application there is also a high degree of variability in the networking context: from a low mobile network of a few nodes to a highly mobile network with thousands of nodes.

This high degree of variability in the networking environment calls for a new design paradigm where network elements (nodes) should be able to adapt to totally different scenarios, engaging in a different behavior depending on the situation. Thus, next generation networks should be able to learn their environment/context and adapt their behavior accordingly in order to achieve their goals. In this paper we introduce some key mechanisms required to enable broad adaptability. Although these mechanisms are general and common to a large variety of tasks/services (e.g. service discovery, location management, cooperative computing, clustering, etc.) we will discuss them in the context of the routing service, leveraging our past experience on the area. This will allow us to ground the discussion in concrete terms and the reader to better visualize the concepts.

Key words: ad hoc, wireless, autonomic, adaptability, routing.

---

[*] This work is funded in part by the EU-funded project ACCA and the EU Network of Excellence E-NEXT.

M. Smirnov (Ed.): WAC 2004, LNCS 3457, pp. 229–244, 2005.

1   Introduction

In traditional networks, network nodes were carefully designed to support specific network capabilities and were deployed and controlled by the owner(s) of the infrastructure. The ever-increasing user population was clearly located at the periphery of these networks and received a service tightly prescribed by the specific network.

Nowadays, the networking landscape is changing dramatically. The numerous network users not only demand new and versatile application support by the networks (e.g., mobility, multimedia, etc) but they themselves are becoming part of the network (network routers, caches, processors, etc) by contributing their resources to it and being engaged in ad hoc networking structures. As the diverse user population becomes more and more part of the networking infrastructure it is clear that networks are bound to be dominated by a new type of network nodes which are much more nomadic, diverse and autonomic.

In the wireless domain, WiFi (802.11-based) networks have opened up the way to high-speed wireless support of autonomic, nomadic users. In addition to the proliferation and enhancement of these networks with numerous extensions (QoS, ad hoc and other capabilities), other ways of networking such nodes requiring higher transmission rates and more demanding application support (mobile and sensor ad hoc networks) have emerged. The resulting rapid deployment of (increasingly autonomic) network elements (nodes) is leading to the formation of large and ever increasing multi-hop wireless networks. Indeed, wireless ad hoc networks of thousands of nodes are already in the designing phase for the USA military's Joint Tactical Radio System (JTRS) Cluster 1 program [1]. However, a large number of nodes coupled with the inherent limits of the end-to-end throughput of ad hoc networks lead to potential bandwidth starvation, unless extra care is taken to develop highly scalable algorithms. For some services/tasks, such as routing in homogenous networks [6], even the best solutions cannot prevent bandwidth starvation when the network size exceeds a certain size (the "curse of dimensionality").

To complicate matters further, the environment a node may encounter may be quite heterogeneous. The scenario a node may encounter and the corresponding best solution may vary greatly, not only in space and time, but also among different nodes sharing the same position (e.g. some nodes may be highly mobile while others may be static, while another pair of nodes may be moving together - i.e. relative mobility is zero). We refer to this as the "curse of diversity".

Finally, the objective of the network may also vary greatly: from allowing peer-to-peer communication among each pair of nodes to detecting moving objects and report it to a central server (sensor networks) to finding the closer provider of a service (e.g. find the closer free parking space) to propagate traffic condition information and negotiate vehicle speeds on an Autonomous Vehicle Network (AVN). Some of these goals are more challenging than others. For example, peer-to-peer communication in an homogeneous network is not scalable with respect to the number of nodes. In the other hand, finding the closer parking lot or locally coordinating among neighboring cars both scale well with network size. Obviously, an application-driven adaptive solution that target the particular task at hand is required, since a general "one-size-fit-all" solution will be inefficient (an overkill) for some practical applications.

In view of the above, it is clear that network nodes are likely to become part of fairly diverse – in characteristics and size – networking environments. The only way nodes belonging to a **L**arge-scale, **A**utonomic, **D**iverse and **A**dhoc **(LADA)** network can cope effectively with such situations is by being able to learn their environment/context and adapt their behavior accordingly in order to achieve their goals (e.g. routing, service advertisement, content distribution, etc.).

In this paper we introduce some key mechanisms required to enable broad adaptability for LADA networks. Although these mechanisms are general, and common to a large variety of tasks/services (e.g. service discovery, cooperative computing, clustering, etc.), we will discuss them in the context of the routing service, leveraging our past experience on the area. This will allow for a more grounded, concrete discussion helping the reader to better visualize the concepts.

This paper is organized as follows. The next section discusses a framework for routing adaptability and introduces two key mechanisms: limited information dissemination (LID) and pattern extraction (PE). The next two sections that follow, discuss these two mechanisms in more detail in the general context of adaptability (i.e. not only for routing). The final section presents some conclusions and suggestions for future research work.

2   A Framework for Multi-mode Routing

Traditional routing protocols for mobile ad hoc networks are usually designed with a particular environment in mind and fail to adapt to the wide range of environments present in an ad hoc network. Because of the wide diversity of the conditions that may be encountered in an ad hoc network it seems that it would be difficult to effectively route information by engaging a single type of protocol. Instead, a multi-mode protocol should be developed which applies the appropriate "mode" or protocol that is determined to be effective at a given point in time and for the appropriate subset of the network. Thus, a multi-mode routing protocol should adapt itself to the present network conditions taking into consideration the traffic levels and patterns (i.e. the application-driven objectives), as well as the mobility patterns (i.e. environment constraints). In order to identify and utilize the network conditions, the multi-mode routing protocol has to rely on some structure-learning/engaging algorithms that extract the network state information (defined in terms of proper metrics) and, based on it, decide on the proper mode to apply to reach each destination.

In [2, 3] the framework for a multi-mode routing protocol shown in Fig. 1 was introduced. This framework proposes that a multi-mode routing protocol – running simultaneously at each node – consists of three elements: two complementing structure-learning/engaging modules that provide network state information

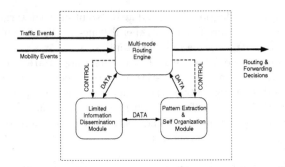

Fig. 1. Multi-mode Routing Protocol Framework

to the third module, the *multi-mode routing engine*, which decides on the mode to apply based on the state of (parts of) the network.

The *multi-mode routing engine* receives information about traffic events (new session requests, or reception of packets to be forwarded to their destination) as well as mobility events (as, for example, nodes displacement and/or link creation/breakage) and passes this information to the structure-learning/engaging modules. Based on this information as well as exchanges among peer modules in neighboring nodes (for example, Link State Update – LSU – messages), the structure-learning/engaging modules obtain some information that defines the state of the network. This information is then passed to the multi-mode routing engine, which uses it to decide the proper routing modes to engage. The behavior of the structure-learning/engaging modules is not fixed but governed by parameters that are defined by the multi-mode routing engine. Thus, the function of the modules is to provide information to the multi-mode routing engine, which controls the modules as well as the final routing mode for each particular packet/destination.

The first of the modules, the *limited information dissemination* module, is responsible for implementing the principle: "the closer you are, the more information you have". This module is in charge of providing detailed information about nodes close by, as well as rough and maybe outdated information about nodes far away. This information may be disseminated in a number of ways. To focus the discussion, LSU messages were chosen as the bearer of the information. This choice was motivated by the fact that link state-based routing presents several desirable properties as for example: fast convergence, well-understood dynamics, loop freedom, etc. However, we should keep in mind that alternatively distance vectors or other metrics (position, service advertisement/description, etc.) may be used as information bearer, and therefore, different algorithms may be executed in the *limited information dissemination* module.

The LID algorithm limits the depth of LSU propagation, avoiding congesting the network with excessive routing overhead in networks with high rate of topological change. Because of the LID algorithm, every node will have good knowledge about the state of its closer links and of far away stable links. This

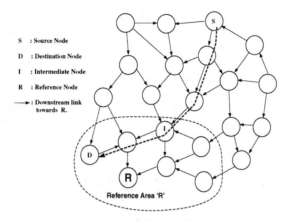

S   : Source Node

D   : Destination Node

I   : Intermediate Node

R   : Reference Node

——▶ : Downstream link
        towards R.

Fig. 2. Routing using the Reference Node/Area (RN/RA) concepts

information will be used by the multi-mode routing protocol to construct links toward close destinations and even to destinations far away in the presence of stable links. When the LID algorithm is applied to a network with a low rate of topological change the result would be the same as if standard link-state algorithm were applied. When the LID algorithm is applied to a network with a high rate of topological change nodes will have detailed information for nodes close to them, without incurring excessive network overhead. This information needs to be combined with some rough information about how to route packets to nodes far away. This information may be provided by some complementary algorithm as the Self-Organizing algorithm discussed below. A family of efficient LID algorithms with good characteristics is presented in Sect. 3.1.

The second structure-learning/engaging module, is the *Pattern Extraction (PE) and Self Organization (SO)* module. The SO algorithm provides an efficient mechanism for reaching destinations far away for which the LID module fails to provide a route. Specifically, the SO algorithm tries to reduce the number of broadcasts required by a route discovery or flooding algorithms by providing pre-calculated routes *toward* some destinations that are *likely* to be involved in new communication sessions. For those routes to be useful, the cost associated with their maintenance should be less than the expected gain of using these routes. The SO algorithm bases its decisions on the traffic as well as mobility patterns of the nodes. It attempts to choose Reference Nodes (RN) and around them Reference Areas (RA) such that the expected number of new sessions having a destination inside the reference area (Gain, G) be maximized. This gain (G) has to be compared against a threshold (the cost of tracking the RNs plus the – hopefully one-time – location management cost) to decide whether it is worth creating routes toward a particular RA. Finally, the SO algorithm either provides information about links toward the RAs (see Fig. 2), or an indication of the highly mobile status of (some of) the destinations.

The gain function used by the SO algorithm is equal to the expected number of broadcasts saved (with respect to reactive route discovery flooding) if the node computing the gain function were to become a RN including (some of) its k-neighbors in its RA. A broadcast will be saved if the destination of a new session is inside a RA. Let node A be a potential RN and let $V(A, t)$ and $G(A, t)$ be the set of k-hop neighbors and the gain function of node A at time t, respectively. Each node $i \in V(A, t)$ has two parameters : $S_i(A, t)$ (probability that node i will stay inside node A's k-hop neighborhood in the immediate future) and $R_i(t)$ (expected number of new sessions having node i as destination in the immediate future). Then, the gain function $G(A, t)$ is defined as:

$$G(A, t) = \sum_{i \in V(A, t)} S_i(A, t) R_i(t)$$

Different approaches can be considered to estimate the values of $S_i(A, t)$ and $R_i(t)$, depending on the desired amount of complexity. We chose the following estimator:

$$\hat{S}_i(A, t) = (1 - \rho) \sum_{j=0}^{\infty} \rho^j Asso(A, i, t - j\Delta t)$$

Where $Asso(A, i, t)$ is a function representing the instant association between nodes A and i at time t, and $0 < \rho < 1$ is the forgetting factor. The forgetting factor determines the extent of the "memory" of the estimator. Larger values of ρ will imply long memory and therefore slow reaction to instantaneous variations. Indeed, the *rising time* (i.e. time required to reach 63% of the desired value when the quantity to estimate is constant) associated with this estimator is $\Delta t/|ln(\rho)|$, which is close to $\Delta t/(1 - \rho)$ for values of ρ close to 1. On the other hand, smaller values of ρ will reduce the memory length and result in a faster reaction to changes, but at the same time it will increase the probability of false alarms (i.e. estimating there is an association between two nodes where there is none). $\hat{S}_i(A, t)$ is easily computed by means of the following recursion:

$$\hat{S}_i(A, t + 1) = \rho \hat{S}_i(A, t) + (1 - \rho) Asso(A, i, t + 1)$$

Thus, to compute the gain function, a node only needs to keep the past value of $\hat{S}_i(A, t)$ and the current value of $Asso(A, i, t + 1)$ for each node i (k-hop neighbor). The association function $Asso(A, i, t)$ is chosen to be a function of the distance $d(A, i)$ between nodes A and i at the current time t, which node A can compute based on the topology information provided by the LID module for nodes close by. If node i is not reachable using node A's topology table entries it is assumed that $d(A, i) = \infty$. Finally, for the estimation of $R_i(t)$, feedback from the upper layer should be employed. Note that the task of estimating these quantities is performed by the so-called Pattern Extraction (PE) sub-module inside the SO module.

If all the nodes are assumed to have the same traffic patterns (i.e. $R_i(t) =$ constant), then the Pattern Extraction (PE) sub-module will attempt to find the mobility pattern of the network. In particular, $\hat{S}_i(A,t)$ represents an estimation of the long-term association of nodes several hops away, as for example the association among nodes in the same group in group mobility scenarios[1]. Although it is possible that the mobility pattern of a network be totally random, that is not usually the case. Human mobility, for example, is based on groups (forming clouds) or follows some patterns (streets, highway, searching, etc.). Even automata mobility is shaped by the function they are executing and therefore there is some degree of spatial/temporal correlation. The self-organizing algorithm will attempt to find (or select) the mobility "leaders" (nodes around which others node move). For example, in networks formed by cars in a highway, the cars in the intermediate position would be the best candidates for mobility "leaders". However, node mobility is not the only factor to take into account. Even more important is the traffic pattern of the nodes. There is no need to pre-calculate routes for nodes that are not going to communicate at all, whereas there maybe other nodes that may need to be contacted frequently due to their mission (coordinator, server, etc.). For the latter nodes it should be highly desirable to have routes readily available saving the network from otherwise almost certain broadcasts. By considering the values of $R_i(t)$ when computing the gain function, the SO behavior is application-driven.

Finally, it was pointed out that a RA will be created only if it is effective. For networks (or some nodes) with high mobility rate or low traffic demand it may not be effective to create them. To forward packets to those nodes route discovery will be used. Similarly, if the routes toward the destination are invalidated too quickly, or if the traffic per session is low to the point that simply flooding the packets is expected to be more effective, then flooding will be used.

Summarizing, thanks to the information provided by these structure learning/engaging modules, each node's multi-mode routing engine will have knowledge of the state of some links (the closer ones and even some links far away that are stable), as well as links towards some regions of the network (RAs) together with information regarding the location (i.e. RA membership) of some destinations. Based on this information the multi-mode routing engine may select its "mode" of operation. Possible decisions include the use of a pre-calculated path of stable links (if available and if stable links are not congested); the use of links toward the destination node's RA expecting that the packet at some point will find a node with knowledge of routes toward the destination as shown in Fig. 2 (this routing mode resembles the Landmark Routing[7, 8] philosophy); the use of a query or a broadcast packet to get the destination node's location information; or simply use a combination of route discovery/flooding.

It should be noted that although the SO algorithm, the creation of reference areas, and the specific "modes" of operation of the multi-mode routing protocol are all particular to the problem of routing, the Limited Information Dissemination (LID) and Pattern Extraction (PE) algorithms have a much broader

---

[1] Since this pattern extraction is the key to the development of adaptable algorithms, it will be further discussed in Sect. 4.

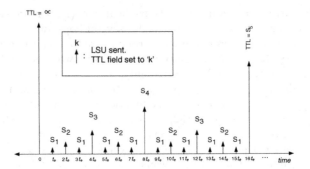

Fig. 3. LSU generation/dissemination process under FSLS

applicability as enabling blocks for self-adaptation for different services/tasks besides routing. We will discuss them in more detail - in the context of broad adaptability - in the next sections.

3   Limited Information Dissemination (LID)

From the start, from the selection of the bootstrapping mechanism to be employed, a node needs some level of awareness of global information. The most basic global information is the network size. For example, employing global advertisement as a bootstrapping mechanism may not be recommendable if the network size is too large. Subsequently, more refined pieces of (global) information may be required in order to make other decisions. The Limited Information Dissemination (LID) module takes care of providing each node with this info.

3.1   A Family of Efficient, Flexible and Adaptable LID Algorithms

In our multi-mode routing approach, a general family of LID techniques referred to as Fuzzy Sighted Link State (FSLS) was employed. The general approach to information dissemination is shown in Fig. 3 (recall that the basic information-bearing element is the Link State Update, or LSU). Every t_e seconds a LSU is sent to nodes up to s_1 hops away, every $2t_e$ seconds a LSU is sent to nodes up to s_2 hops away, each $4t_e$ seconds a LSU is sent to nodes up to s_3 hops away, and so forth. The values of the $\{s_i\}$ sequence depends on the scenario. For example, for the case of routing for flat homogeneous networks, it was shown in [4] that the optimal sequence (referred to as HSLS) was $s_i = 2^i$. Not only that, but employing HSLS achieves the best scalability properties among known routing protocols. Indeed, that result showed that having different levels of information awareness (from fine grained local information to low resolution global information) pays off.

HSLS's effectiveness is due to its exploitation of the locality of the effect of most link changes. Indeed, for hop-by-hop routing, the extent of a node's decision is limited to choosing the best next hop for a path among its one-hop neighbors. It turns out, that for large scale networks and destinations far away, the probability of making a good next hop decision is related to the angular displacement of the destination with respect to the node making the next hop decision. For example, if the node thinks that the destination is in the "North" direction, it will relay packets to this destination to its northernmost neighbor, and this decision will remain valid as long as the destination remains "on the north". Since this angular displacement depends on the ratio between node movement since last update (latency time times node speed) and the distance to this node, we conclude that the probability of making a bad next hop decision depend on the ratio between latency-of-link-state-information and distance. For uniformly distributed traffic, the best solution is obtained by keeping this probability of mistake bounded (almost constant) and this results in the HSLS schedule.

However, if the traffic distribution is not uniform, e.g. the traffic tends to be localized, different schedules can be considered. One particular schedule that is useful when the traffic is localized - as well as in order to provide pertinent information to the pattern extraction (PE) module - is the Near Sighted Link State (NSLS) schedule. In NSLS, all the event-driven LSUs are distributed only to nodes at a distance of k hops or less. That is, in NSLS $s_i = k$ for all k. These LSUs are complemented by periodic (long interval, seldom sent) global LSUs. Thus, the particular schedule to use for the $\{s_i\}$ sequence will be determined - among other things - by the range of impact of information changes (link state changes in the case of routing), the expected (average) cost of the mistakes induced by the inaccurate information, and the requirements of the PE module.

Among the pieces of information that can be extracted from a topology table filled by a LID module, even when the data is outdated, are:

- A rough estimate of the network size. The network size, which is not supposed to change frequently or dramatically, is basically needed to make the small/large network classification and decide on the methods (modes) of operation. For example, if the network is detected to be large, then global flooding should be avoided as a service advertisement mechanism.
- A Close/Far classification for each destination. This information is useful to determine the mode to engage for each.
- A Sparse/Dense classification of the network. If the network is regarded as dense, corrective actions need to be taken (e.g. topology control, use of multi-point relays, setting of parameters at the MAC layer, etc.).
- A Slow/Fast moving classification for each destination, based on the variation of its distance to other nodes over time.
- Provide the Pattern Extraction module with the information necessary to estimate the degree of association between this node and the other nodes in the network (see next section).
- For the routing service, provides the ID of a next hop towards each destination. Depending of the schedule of LSUs employed, this next hop decision may have a high probability of success.

3.2   Some Adaptation Services Enabled by a LID Mechanism

As it was explained before, LID is not limited to bearing link-state information, or to the routing service. For example, if geo-location information is available, the node position can be advertised instead of the LSU. Reactive routing protocols can then send Route REQuests (RREQ) to a specific region of the network where the destination is highly likely to be located, instead of sending the RREQ packets to the entire network. Another example is service advertisement. It was already pointed out that a service advertisement approach based on nodes sending global advertisement would not work for large networks, since they will consume most of the bandwidth in such advertisements due to the broadcast storm problem. Thus, a more sensible approach is to send local advertisement and to progressively increase the depth of propagation of such messages. This way, upon bootstrap a node will advertise its services to nodes, say, 2 hops away. It will wait some time and re-send the advertisement to nodes 4 hops away, and so forth. While doing so, since it will also be receiving advertisements from other nodes it will start to learn how big the network is and will adjust its timer/advertising schedule accordingly. Eventually, all the nodes in the network will learn of the server after a time proportional to the network size and its distance to the server, avoiding having the network collapse due to a broadcast storm during initialization.

Another example of the applicability of LID techniques can be found in the area of dynamic spectrum allocation, where a node's transmission frequency is not fixed beforehand, but it is computed on-the-fly based on sensing of the environment and the regulatory policy in effect. Such nodes are being under development as part of the DARPA's neXt Generation (XG) program[5]. Once the nodes have estimated the characteristics of the electromagnetic spectrum around them (e.g. the presence of incumbent nodes with primary rights over a particular bandwidth in that area) they will try to schedule (both in time and in frequency) their transmissions. To this end, nodes will require detailed information about nodes close by (say, up to 2 hops away) to perform distributed MAC scheduling algorithms. At the same time though, the nodes may greatly benefit from loose information about spectrum availability of nodes far away. This loose information (lower granularity, dividing the frequency in big chunks of spectrum and reporting aggregate usage over them) will be used to find "sweet spots" of global spectrum availability. These sweet spots will be the best candidates to look for a global coordination channel that all nodes are tuned to, and therefore can be used to broadcast packets to all nodes in transmission range, not only to those nodes *known* to be neighbors. Such a coordination channel is necessary to perform functionalities such as neighbor discovery/link setup and link maintenance (e.g. when a link becomes invalidated due to the arrival of a primary node with exclusive rights of usage over parts of the link's spectrum).

As we may see, LID is not only useful for routing, but it is a more general design paradigm that is neither local nor global. The basic tenet of it is that *some* global information is better than none. Designing algorithms based only on local information is equivalent to trying to find your way out of a forest by walking (and watching) at the ground level. Using traditional centralized (or decentralized) algorithm where full global information is available is equivalent to climbing the highest peak available to figure out the way out of the forest. Using FSLS is similar to the sensible approach of climbing, from time to time, a small hill to get a sense of the surroundings and make the decision on the path to follow next, until the next hill. Note that the height of the hill that need to be climbed may also depend of the scenario (e.g. height of the trees) and therefore the FSLS must be adaptive to the scenario (feedback control loop).

We can easily see how the LID concept can be useful for a variety of autonomic systems. Take for example the case of an automatic vehicle network. A vehicle will need to exchange detailed information with vehicles close by in order to - among other things - avoid crashing with them. However, as the distance to the vehicles increases, only rough pieces of information are needed. For example, a vehicle may only be interested in the total number of nodes far away and their average speed, as to determine the likeliness of a traffic jam. Once again, the specifics of the information being propagated, the algorithms being run over the data and the best information propagation dissemination schedule is dependent on the task at hand. For flat routing, the best solution can be found in [4].

4   Pattern Extraction

Whether patterns are observable in a network or not, depends on the scale (space/time) we use to observe it. For example, let's consider a network formed by pedestrians and cars in a city. If we zoom out and see the network from the outer space, the entire network would look as a single group, well contained inside the city boundaries. On the other hand, if we look at the network from the ground, as seen by a pedestrian user with a very limited transmission range, the network would appear totally chaotic, with new neighbors appearing and disappearing. Obviously, both observations (city level and pedestrian level) would be of little use. However, a more sensible approach would be to look at the network as seen from a tall building. In this case, we could observe the different mobility patterns induced by the streets and nodes moving to similar destinations. We'll notice some cars trying to get out of the city and some trying to get in. We can group the cars according to their direction even though they momentarily move apart due to traffic conditions (traffic lights, etc.). The same observation can be made about the timescales. Cars whose trajectory may appear to have no connection while observed at a small timescale, may be discovered to be headed to a similar destination if observed over a longer period of time.

Thus, patterns are present in most networks. They are typically induced by the environment they operate in (e.g. cars on a highway, humans in a university, etc.) or the mission they fulfill (e.g. robots helping in disaster recovery operation, self-deploying sensors, etc.). In order to detect the useful patterns, we need to look beyond the one-hop neighborhood and determine the observation frequency(ies); that is, observe at the proper time/space scale.

While it may be easy to agree that patterns are present, the main question is *how hard are they to find*; that is, *will finding patterns be computationally feasible?* We answered this question for the case of routing for ad hoc networks under group mobility. We applied the associativity estimator $\hat{S}_i(A,t)$ presented in Sect. 2 to detect the group leaders in networks when group mobility is present. It should be noted that computing the estimator for a set of nodes k hops away ($k = 2$ in our experiments) did not incur significant processing overhead, since thanks to our recursive expression we basically needed to keep one place in memory for the estimator of each of the k-hop neighbors and run the update operation using the information about the nodes' current distance provided (as a by-product) by the LID module (that was computing the shortest path first tree for all the destinations). Thus, our estimator - while not optimal - showed the feasibility of detecting patterns at a low computational cost.

In Fig. 4 we show the results of applying the estimator to a 100-node network consisting of 5 groups. The estimator was used to determine the gain function (see Sect. 2 and [3]) of each node and choose the best candidates to become RNs (equivalent to cluster leaders). Ideally, the group leaders would be elected RNs and there should be as many RAs (equivalent to clusters) as mobility groups. In practice, however, estimation is not perfect. This is illustrated in Fig. 4, where the curve MoI (mobility only, ideal) represents the situation where the RNs are chosen based on the mobility pattern only, and they are chosen to be exactly the group leaders (ideal situation, where the identity of the group leaders is known beforehand). The curve MoC (mobility only, computed) represent the corresponding situation where only mobility patterns are taken into account and the estimator $\hat{S}_i(A,t)$ is used to select the best candidates to RNs. We can see that in general our estimation is not perfect and we loose some performance with respect to the ideal case, but still we obtain a good solution at a reasonable (computational) cost. It is interesting to note that there is a particular case (low mobility) when the estimator failure to properly detect the group leaders actually improves performance. This is due to the fact that during the simulation lifespan two groups were close to each other giving the impression of being only one. The estimator chose one node in the intersection of these two groups as the RN. And, since there was slow mobility, during the simulation lifespan the two groups acted as one. Choosing the wrong group leader actually helped performance. After a while, though, after the groups grew apart, there would have been a penalty for the bad selection, although it may not be big enough to counter-balance the gain from grouping the two sets of nodes together for a long period of time. This once again raises two important points: (i) timescales are important when detecting patterns, and (ii) we shouldn't loose sight that our goal/objective is not to find the groups/group leaders but to effectively deliver packets to their destinations. Our gain function captures this goal (application-driven approach). Thus, what really matters is that − at the routing protocol timescale − the nodes present a pattern that can be exploited for effective data delivery, even though in the long run the pattern may not hold.

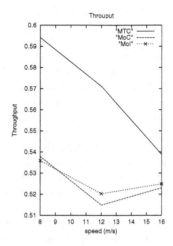

Fig. 4. Throughput for a 100 node network under different speeds using different estimators for choosing the reference nodes

Finally, the remaining curve, MTC (mobility and traffic) represents the case when both mobility as well as traffic patterns are used. In this set of simulations, not all the nodes were destinations of active flows. Only a small group of nodes were actually involved in communications and they could successfully be grouped in two RAs. Thus, applying the application-driven self-organizing criteria that RAs are to be created only when it pays off, the SO algorithm decided to build/maintain 2 RAs only, reducing the protocol overhead and increasing the throughput. The simulation results show the evident superiority of considering the traffic patterns above mobility patterns.

The results shown in Fig. 4 refer to a particular system addressing the issue of multi-mode routing, and some specifics of the solution (gain function, reference area creation/tracking) are relevant only to the particular solution to the routing problem. The Pattern Extraction (PE) sub-module and the estimation being used, however, are much more general and applicable to a wide variety of problems. Basically, the key element to the success of the routing solution was the ability to detect associations between nodes at a bigger time scale than the immediate present and at a bigger space scale than the one-hop neighborhood. We were able to detect associations between nodes 2 or more hops away (as for example cars moving towards the same destination on a highway, where sometimes one of them passes the other and moves 2 or 3 cars away until the other ones catch up, and vice versa) and for a longer time period, of the order of the routing events. These associations, when present, form the backbone of the network. Thus, extracting patterns means to move beyond the search for stable links (one-hop associations) and start looking for stable associations at longer distances/time scales. These associations can then be exploited in a lot of different ways, depending on the task on hand. For example:

- In reactive routing techniques, we may prefer to choose paths where a subset of the nodes present strong associations between them. For example, let's assume that in the route $S - a - b - c - d - e - f - g - D$ nodes S and c present a strong association, as do nodes c and f, and nodes f and D. Thus, we may refer to nodes S, c, f, and D as "anchor" nodes since we may use them as anchors to maintain the route from S to D. For example, in case of a link breakage in the segment $S - a - b - c$, node S does not need to initiate a global repair of the route, but since node S knows that node c is likely to still be around, node S may issue a local request to built a new path to node c (instead of to the destination D). Once the path from S to c is repaired, with say a new segment $S - a' - b' - c$, the path to the destination becomes $S - a' - b' - c - d - e - f - g - D$. Thus, the use of anchor nodes (and pattern extraction) allows us to avoid global route repair localizing the effect of link breakage (a scalable approach).
- Knowledge of the underlying network patterns and backbone help to build stable structures, as for example grouping nodes in long-lived/stable clusters (or, as in our routing example, the reference areas). Reducing the instability of the structures built on top of a network (e.g., clusters) significantly reduces the overhead needed for repair/maintenance.
- Knowledge of patterns help to classify nodes according to their characteristics and to determine appropriate modes of operation for each node/region.

It can be seen that finding associations can significantly improve performance. However, it should be noted that finding these associations requires non-negligible time. This is true because, as mentioned earlier, the timescales at which we look (and care) for patterns is the same as the timescale of the application that will exploit the patterns (routing, in our experiments - case study). Thus, the convergence time for the estimator will be of the same order of magnitude of the network time scale, which is typically not small. For example, it will take a time in the order of $\Delta t/(1 - \rho)$ seconds for the estimator $\hat{S}_i(A, t)$ (presented in Sect. 2) in our routing experiments to discover associations. The value of ρ cannot be too small since it this case the probability of false alarm will increase significantly and the structures formed based on the misunderstood patterns will not be stable (at the network time scale). As a consequence, initialization (initial discovery of patterns) will take a time in the order of the network timescales. Since this value is typically long, alternative techniques need to be used to provide service during this long initialization period. For example, flooding Route REQuest can be employed in the routing example to provide service to destinations until the pattern-based structures are present. Note that attempting to reduce this initial convergence time is likely to lower the quality of the estimator and decrease the network performance in the long run. Also, if the network patterns as observed at the network timescale start changing, it will take a non negligible time for the PE module to track these changes. In our experiments with networks exhibiting group mobility under realistic scenarios (speed, transmission range, etc.) corresponding to vehicles moving on the ground, it took several hundred seconds before patterns could be extracted and reference areas could be formed. In the meantime, the multi-mode engine acted as if no pattern where present (worst case) resorting to outdated link state info or route request (depending of the destination) to deliver the packets.

5   Summary and Concluding Comments

As elaborated on earlier, network users not only demand new and versatile application support by the networks but they themselves are becoming part of the network. These users are becoming much more nomadic, diverse and autonomic, creating a fairly diverse in size and characteristics networking environment. Consequently, it is clear that future networks will increasingly:

- have ad hoc, changing and rather large structures,
- be designed for operation in diverse environments (heterogeneous) and
- consist of network nodes that will be both diverse and autonomic.

As a result, network nodes are likely to find themselves in (as well as contribute to shaping) large, ad hoc and quite diverse (LADA) networking environments. Unless carefully designed, the formed LADA networks would suffer greatly from the curse of dimensionality and diversity (and possibly an emerging one, the "curse of autonomicity") and would be fairly inefficient if functional at all. Autonomicity may be viewed as not only simply capturing a changing or variable behavior of nodes, but also autonomous behaviors that may be random or shaped by tasks, rules, policies or the environment.

Mechanisms that sense the network conditions and take decisions / adjust key parameters, have long been around (e.g., ethernet, etc). These mechanisms utilize some network state information (that they extract themselves or are being provided) and adjust their behavior / parameters. These "basic" adaptation mechanisms are completely inadequate for the large scale, ad hoc and heterogeneous LADA networking structures, composed by autonomic and diverse network nodes. Coping effectively with autonomicity, diversity and dimensionality that are inherently affecting the emerging networks requires a more comprehensive approach to adaptation than the "basic" one of the past.

Information dissemination is the cornerstone of an effective adaptable LADA network. Not only it will have to overcome the curse of dimensionality itself but also provide adequate information to enable the deployment of scalable (i.e., cope with the curse of dimensionality) and effective (multi-mode) network protocols. To facilitate the latter, the disseminated information should be sufficient to help other nodes **characterize the collectively shaped networking environment** as well as for **extracting behavioral and other patterns** (i.e., cope with the curse of autonomicity and diversity) in the network and feed adequately **self-organizing mechanisms**.

This paper has demonstrated the aforementioned advocated approach for LADA networks by applying it to large scale, adaptable, mobile, ad hoc networks. The effective operation of (autonomic) nodes in ad hoc networking environments relies strongly on their ability to adapt to it; that is, learn about

244 C.A. Santivanez and I. Stavrakakis

the specific environment and invoke the appropriate protocols. To enable this adaptation, a (scalable) Limited Information Dissemination algorithm is necessary to provide to the nodes local and global network information of various resolution levels. This information will be processed by the nodes, so that they detect key characteristics of the environment, possible organize themselves and finally invoke the proper protocol functionality. The aforementioned approach has been outlined (and applied successfully) for adaptive (multi-mode) routing in ad hoc networks, where the collected local and global information is processed by the Pattern Extraction and Self Organization algorithms. Other examples of adaptation services that require a Limited Information Dissemination protocol (such as dynamic spectrum allocation) are also presented (Sect. 3.2).

In addition to the framework for managing LADA networks proposed in this paper, this paper provides some ideas and algorithms for implementing specific functionalities of this framework, as well as attempts to bring out key issues that should be addressed to enable the proposed framework both for routing as well as other service support in LADA networks. Such issues include (scalable) modulation of the disseminated information in space and time to achieve diverse resolution for local and global information (LID algorithm), rules for information compression/aggregation/merging for scalable support of services other than routing, consideration of the appropriate time-scales for extracting behavioral and other patterns as well as algorithms to deliver them, rules for self-organization for scalable service provision, stability considerations of the adaptation strategies, etc.

References

1. The Joint Tactical Radio System (JTRS) program web page at http://jtrs.army.mil/
2. Santivanez, C., Stavrakakis, I.: A Framework for a Multi-mode Routing Protocol for (MANET) Networks. Proceedings of IEEE WCNC'99, New Orleans, LO, September 1999.
3. Santivanez, C.: A framework for multi-mode routing in wireless ad hoc networks: theoretical and practical aspects of scalability and dynamic adaptation to varying network size, traffic and mobility patterns. Doctoral thesis, Electrical and Computing Engineering Department, Northeastern University, Boston, MA, November 2001.
4. Santivanez, C., Ramanathan, S., Stavrakakis, I.: Making Link State Routing Scale for Ad Hoc Networks. Proceedings of MobiHOC'2001, Long Beach, CA, Oct. 2001.
5. DARPA's neXt Generation program: web page http://www.darpa.mil/ato/programs/XG/
6. Santivanez, C., McDonald, A. B., Stavrakakis I., Ramanathan, S.: On the Scalability of Ad Hoc Routing Protocols. Proceedings of IEEE Infocom'2002, New York, USA, June 2002.
7. P. F. Tsuchiya, P.F.: Landmark Routing: Architecture, Algorithms, and Issues. Technical Report MTR-87W00174, Cambridge, MA, MITRE Corporation, September 1987.
8. Pei, G., Gerla, M., Hong, X.: LANMAR: Landmark Routing for Large Scale Wireless Ad Hoc Networks with Group Mobility. Proceedings of ACM Workshop on Mobile and Ad Hoc Networking and Computing MobiHOC'00, Boston, MA, August 2000.

BIONETS: BIO-inspired NExt generaTion networkS

Iacopo Carreras, Imrich Chlamtac, Hagen Woesner, and Csaba Kiraly

CREATE-NET, Trento, Italy
{carreras, woesner, chlamtac, kiraly}@create-net.it
http://www.create-net.it

Abstract. The amount of information in the new emerging all-embracing pervasive environments will be enormous. Current Internet protocol conceived almost forty years ago, were never planned for these emerging pervasive environments. The communications requirements placed by these protocols on the low cost sensor and tag nodes are in direct contradiction to the fundamental goals if these nodes, being small, inexpensive and maintenance free. This situation needs therefore a radically different approach to communication in these systems, especially since pervasive and ubiquitous networks are expected to be the key drivers of the all encompassing Internet of the coming decades. The fundamental disparity between the need for extremely dispensable, low cost devices, such as sensors or tags, and increasing communications load per device due to the presence of billions of nodes, that is creating an unbridgeable paradox, is therefore an insurmountable obstacle on the way to adoption when conventional networking architectures are being considered. Biological systems provide insights into principles which can be adopted to completely redefine the basic concepts of control, structure, interaction and function of the emerging pervasive environments. The study of the rules of genetics and evolution combined with mobility, leads to the definition of service oriented communication systems which are autonomous, and autonomously self-adaptive. The objective of this article is to ascertain how this paradigm shift, which views a network only as a randomly self-organizing by-product of a collection of self-optimizing services, may become the enabler of the new world of omnipresent low cost pervasive environments of the future.

1 Introduction

As the trend toward ubiquitous and pervasive computing continues to gain momentum, the number of nodes is expected to grow by multiple orders of magnitude as tags, sensors, body networks and myriad of other miniaturized devices get fully integrated into the global communication superstructure. Not only will the amount of information in these all-embracing pervasive environments be enormous and to a large degree localized, but also the ambiance within which these nodes will act will be intelligent, mobile, self- cognitive, knowledge based,

M. Smirnov (Ed.): WAC 2004, LNCS 3457, pp. 245–252, 2005.

and, in a sense, "almost alive". Current Internet protocols, conceived almost forty years ago, were never planned for these emerging pervasive environments. Besides their many limitations with respect to agility, mobility, scalability, etc., they are also too heavyweight in the computing and communication requirements they impose on the small and energy- limited nodes. Hence a totally new way of controlling these emerging systems is needed. Realizing that the raison d'etre for the protocol layers is the support of better services to the user, we consider an approach to the pervasive environments that is derived from the network's original goal - services. Given, as observed above, that the pervasive systems are dynamic, growing, self aware and evolving, it is natural to model them as biological entities. Considering the behavior and hence control of these pervasive environments an evolving organism leads us to applying the rules of combination of genetic material and evolution to determine the information exchange, fully replacing the concept of end to end Internet oriented protocols. By information exchange occurring in these systems only as an individual need of each entity in the biosystem to reproduce, the rules of control become localized, minimized and occur only when and where needed, making it possible for the individual nodes to work with minimal energy, with computing or communication effort expended only as it benefits their own goals. Even more important, by defining themselves as the end result of the evolution process, the services here become characterized by their ability to mutate and select the fittest to survive, this way constantly evolving and self-optimizing leading to a new concept of self aware, self optimizing bionetwork, a candidate for the new world of omnipresent low cost pervasive environments of the future. This article expands on our previous work published in [1] in defining the processes and mapping of the biological approach to pervasive systems communication.

2   Autonomic Pervasive Environment Today

In the new emerging pervasive environments the rules of the game are drastically changing. In previous systems, mobility was something that needed to be handled in order to extend the already existing services to a wireless context. Nowadays, almost every person has a cell phone and almost every person is using the wireless devices for more and more purposes which are not just phone calls. Services are accessed from remote locations, while moving and without a reliable end-to-end connection. In such a new *moving* environment, it is natural to start asking which would be the best way to benefit from this implicit user mobility and how to maximally exploit it.

In the traditional Internet the flow of information follows the source-to-destination philosophy. Within the last years two main research directions evolved that cover the world of pervasive networking: Mobile ad-hoc networks (MANETs) and Wireless Sensor Networks (WSN). While the MANETs try to cope with mobility, the research in WSN concentrated on dealing with huge numbers of power-constraint data sources. Using sensor nodes for relaying packet data from other sensors turned out to be a serious power drain, limiting the life time of sensors. In [2] [3] the authors introduce hierarchy into previously flat sensor network architectures. Here, the users carry data packets collected from sensors or other users and drop it whenever they encounter a gateway to the backbone. This way mobility can be exploited to provide connectivity in a disconnected environment. The information is treated exactly as in the flat networks, but with a different routing policy. Services are the same as well as the entire protocol stack. The same idea of using mobility to create connectivity was shown for MANETs in [4]. In [5] the authors show that mobility indeed increases the capacity of MANETs.

All the previously described networks deal with the transport of data packets, and while there are techniques to do *data fusion* in WSN, their aim is to reduce the number of data packets to forward by eliminating redundancy.

2.1   Network Original Goal Is Service

Trying to find a better solution to the exposed problem this article totally changes the perspective. In the new ubiquitous context the information that is managed and exchanged by users is drastically changing in its significance. Information will not only be to a high degree localized, but also aging, which means that most of the time information will simply be outdated and therefore useless with respect to the context where the user is moving in. In this sense, source-to-destination data transport will be needed in selected cases only (e.g., when the information needs to reach destinations such as remote servers, for permanently storing the gathered information).

Since the service is the original goal of the network, we let the service itself define how the network is supposed to be in order to satisfy its requirements. Networking will occur only as a consequence of service needs and the network itself will evolve and adapt together with the service. In the envisioned ubiquitous scenario the environment will dictate the principal rules of adaption. Users will be mobile and will change their location in short periods of time, leading to a continuously changing environment. The success of the service will be in following these changes and in subsequently adapting its main functionalities. In this sense the network may be interpreted as the *habitat* where organisms are moving and the genetic information codes their behavior and goals.

3   BIONETS

3.1   Related Work

Several examples are available in the literature, where biological concepts are considered as models to imitate. Each one of this examples focuses on a different biological aspect and apply it to solve or to optimize a specific technological problem.

In [6] [7] the swarm intelligence of *social insects* has inspired an evolutionary framework capable of connecting heterogeneous objects and services. This distributed framework should exploit the decentralized organization of autonomous biological individuals leading to an *emergent behavior*. According to these principles in [8] the described framework has been implemented and evaluated in terms of messages delivery efficiency and high workload situations.

In [9] autonomic techniques are applied to system management, where a centralized management approach is unlikely to be adopted due to the enormous number of devices participating in large-scale pervasive scenarios. Drawing inspiration from nature, problems with real-world relevance are approached. For example, in *Flyphones* the autonomic creation of cells macro-structure in animals cells is used to determine the channel allocation algorithm in a mobile telephone network, while a bacterium- inspired software has been developed for the adaptive management of active service networks.

In the above mentioned examples the genetic approach is not considered. Since in nature the genes are the coding blocks of an unbounded set of self-organizing behaviors, this is what we would like to reproduce through the BIONETS approach. Different coded behaviors can then be naturally chosen by the natural selection.

3.2   BIONETS Principles

In the depicted scenario services are associated with living organisms. Service is defined by chromosomes. In this way service evolves and adapts to the environment constantly and autonomously, Chromosomes are collections of *genes* that are the smallest service (related) data unit and inborn intelligence/instincts and thus represent all the information needed for the organism to function and service to be executed.

As in nature, it is possible to define a complete life-cycle of the organisms and therefore of services. The life cycle starts from the birth of an organism, goes through the reproduction and ends with the death. Each one of this stages will be defined in the following. Reproduction and evolution occur applying evolution rules inherited from nature.

Fitness is measuring the correspondence of the organism genetic information with the environment and determines the exchange of information(genetic information). Therefore no end to end communication concept exists in these systems, information is only exchanged as needed, locally, between mating organisms. Environment is determining the *Natural Selection* based on the *fitness* of the organisms with the environment leading to the best services possible as a function of the environment.

The Service Is the Organism. We are envisioning a scenario where users will be more and more interested in a service able to provide reliable localized information. The role of the service will be for instance to provide answers to questions like *How is the weather around the train station?* or *Where will I find a free parking space around there?*. Services will be hosted on users' devices and will go around through the physical movement of the users.

Each service is constituted by a program and its related data that is organized into chromosomes.

The Chromosome. The service will use his genetic information to run its main functionalities. The genetic information is stored in the chromosome. Each chromosome consists of:

- Data that is the genetic information of the organism, organized in genes
- A *plugin* that stores a syntax notation describing the actions dictated by the chromosome and the fitness (degree of attraction) operator yielding *natural selection* through preferred mating

The Gene. Genes are a tuple of information and consist of:

- value
- timing information
- information source ID / location information
- other data depending on the service

Organisms are diploid, meaning that there will always be 2 homologous chromosomes associated to each service. The two homologous chromosomes will have the same genes but in different forms. Alleles are different forms of the same gene and corresponds to different values in the tuple of gene information. They may differ in timing information or in the source value information. Each allele may be dominant or recessive depending on the service behavior.

Having two chromosomes allows us to estimate the reliability of the data. We would probably always choose the youngest data value to be the actual one if it is a parking lot, but we might also average the sensor data if it represents temperature. The choice of the preferred value among the two reflects the concept of dominant and recessive genes.

As in nature, recessive information will enable the service to survive in different environments, providing the service with higher resilience against data corruption and fraud and may even allow for additional features.

3.3   Service Life Cycle

A service is born when the user downloads the chromosome onto his device. From that moment on, the user is able to interact with the other organisms (i.e. users carrying chromosomes) and with the environment where the users are physically moving. When gathering information from the environment the service *grows*. While growing, the service improves its functionalities, meaning that it becomes able to increase performance.

When a user meets another user while moving, services may reproduce and produce offsprings. It is in this phase that evolution and natural selection occur. In order to be able to reproduce the service must satisfy some fitness requirements, since it is not willing to spread useless information. We assume a service to be dead when he can not reproduce anymore.

Birth. The service is born when the user gets (downloads) an empty chromosome of a certain service that consists only of the plugin. From now on the user can read data from sensors and use the syntax definition from the plugin. Note that the user is haploid now, i.e. it has only one chromosome per service.

Growth. When reading sensor data, the user fills the chromosomes both at the same time. This information is in any case more reliable than the previous information on the chromosomes.

Reproduction. The concept of mating is performed the following way: In *Meiosis*, the diploid cell splits into 2 haploid reproductive cells (the eggs or sperms)[1]. This means that the chromosome pair is being split and copied. Two packets are sent out one after another containing one chromosome each. In the best case when all the sent packets reach the receiver, it has 4 different combinations of the chromosome pairs. Note that this is the best case, the user may for energy reasons decide to send out only one chromosome, or receive only one because of packet loss.

A selection has to take place to decide which of these combination survives. This selection could be influenced by the quality of the plugin (by the version number). Having such a selection can help to spread new versions of plugins. It also may help to repair broken chromosomes (that were damaged during the wireless transmission). In a sense, we allow for truly spontaneous mutations and we may even find that mutations spread. It remains to be seen if this is any good.

The selection occurs as a consequence of localization and age. We can define the fitness of a chromosome as the average of the timing information of the genes weighted with the localization information. In this sense the environment participates in the selection, since the survival of a service also depends from where the user is (but not only) when he mate with another user.

Death. If the sensor data in the chromosome is too old, it is very likely useless. Thus we forbid to send out the chromosome after a certain threshold (coded in the plugin). This way the chromosome can die, but it may be reborn through reading new sensor data or receiving a fresh chromosome.

The service is considered *alive* as long as it is able to answer to questions. Death is therefore a consequence of outdated chromosomal information (sensor gathered information is aging). It is in the interest of the user to exchange information and to gather sensor information and this same interest drives the service *instinct to survive*. We have now defined a complete life cycle of the service.

---

[1] In real life 4 haploid cells would appear because of one step of copying the chromosomes including cross-over. Each two copies would be identical.

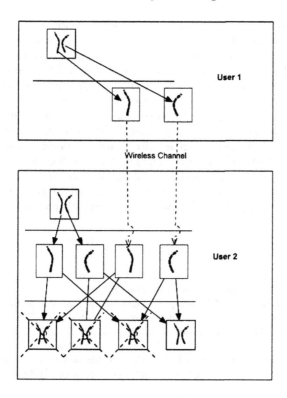

Fig. 1. Mating process

4   Evolution of the Service

The basic principles described in chapter 3, together with the rules of the selection determine the evolution of the service. In the model we have identified the issue of the *instinct to survive*, which is however is still an open question.

Since the service's aim is to provide a measurable *benefit* to the user, this benefit should be the metric to determine the fitness of a certain service. Thus, the simplest fitness criterion is: "User likes the service". This opinion could be measured by explicit user feedback, but other ways of inferring it from the user's behavior should be found.

We are expecting the service to carry some memory on what has determined his successfulness. Hence, the chromosomes will contain not only raw data, changing continuously but also some coded instructions or intelligence.

5   Conclusions and Future Work

We have presented a new concept of information exchange in pervasive networks. The major conceptual shift is the use of networks of occasional information exchanges between mobile users, helping to spread information rather than forwarding data packets. The use of genetic models leads to a population of service instances on a set of user nodes. This population can grow if the service is successful or decline if it is not. In addition to the growth of populations BIONETS allow for the evolution of the service itself through mutations and selection of the fittest. The specific fitness criteria are still to be elaborated.

In order to examine the behavior of the proposed model in a day-to-day scenario we consider the case of a parking lot application. In such scenario the city is split into blocks and each gene will contain the information on the status of a parking spot (FREE or OCCUPIED). The service will guide the users towards the nearest free parking place and will evolve according to the environment where the users are moving. Since first simulations indicated the great potential of this idea, we are expanding the model towards more realistic user behavior right now.

References

1. Chlamtac, I., Carreras, I., Woesner, H. In: From Internets to BIONETS: Biological Kinetic Service Oriented Networks. Springer Science (2005) 75–95
2. R.C. Shah, S. Roy, S.J., Brunette, W.: Data mules: modeling a three-tier architecture for sparse sensor networks. In: Proceedings of the IEEE Workshop on Sensor Network Protocols and Applications (SNPA). (2003) 30–41
3. Zhao, W., Ammar, M., Zegura, E.: A message ferrying approach for data delivery in sparse mobile ad hoc networks. In: Proceedings of the 5th ACM international symposium on Mobile ad hoc networking and computing, ACM Press (2004) 187–198
4. Vahdat, A., Becker, D.: Epidemic routing for partially connected ad hoc networks. Technical Report CS-200006, Duke University (2000)
5. Grossglauser, M., Tse, D.N.C.: Mobility increases the capacity of ad hoc wireless networks. IEEE/ACM Trans. Netw. **10** (2002) 477–486
6. Nakano, T., Suda, T.: Adaptive and evolvable network services. In Kalyanmoy Deb, e.a., ed.: Genetic and Evolutionary Computation GECCO 2004. Volume 3102/2004., Springer-Verlag Heidelberg (2004) 151–162
7. Wang, M., Suda, T.: The bio-networking architecture: A biologically inspired approach to the design of scalable, adaptive, and survivable/available network applications. In: SAINT. (2001) 43–
8. Suzuki, J., Suda, T.: Design and implementation of an scalable infrastructure for autonomous adaptive agents. In: Proc. of the 15th IASTED International Conference on Parallel and Distributed Computing and Systems. (2003)
9. Shackleton, M., Saffre, F., Tateson, R., Bonsma, E., Roadknight, C.: Autonomic computing for pervasive ict a whole-system perspective. BT Technology Journal **3** (2004)

Dynamics, Information and Control in Physical Systems*

Alexander L. Fradkov

Institute for Problems of Mechanical Engineering of RAS,
61 Bolshoy Ave. V.O., St.Petersburg, 199178, Russia
alf@control.ipme.ru

Abstract. The subject and methodology of an emerging field related to physics, control theory and indormation theory are outlined. The paradigm of cybernetical physics as studying physical systems by cybernetical means is discussed. Examples of transformation laws describing excitability properties of dissipative and bistable systems are presented. A possibility of application to analysis and design of information transmission systems and complex networks is discussed.

1 Introduction

Modern communication networks are evolving very rapidly. Their element sizes are going down while signal rates are going up. Among different tendencies influencing development of the principles of the future generation communication networks the following two seem very important.

1. Complexity of systems and networks of the future requires very high speed of information exchange, close to physical limits. It makes conventional study of information flows in the network not sufficient. **Networks should be treated as physical systems with their properties and limitations.** Particularly, energy exchange flows should be taken into account. It means that information should be treated as a physical quantity, like energy or entropy. Quoting R. Landauer, "Information is tied to a physical representation and therefore to restrictions and possibilities related to laws of physics" [1].

An importance of understanding interrelations between energy exchange and information transmission was recognized still in the 1940s. The founder of the information theory C. Shannon derived in 1948 that at least $kT\ln 2$ units of energy is needed to transmit a unit (1 bit) of information in a linear channel with additive noise [2], where T is absolute temperature and k is the Boltzmann constant. This is just an energy required to make a signal distinguishable above the thermal background. In 1949 J. von Neumann extended that statement to the following principle: any computing device, natural or artificial must dissipate

---

* Supported by Russian Foundation of Basic Research (grant RFBR 02-01-00765) and Complex Program of the Presidium of RAS #19 "Control of mechanical systems", project 1.4.

at least $kT \ln 2$ energy per elementary transmission of 1 unit of information [3]. Next contribution into linking energy and information was made by R.Landauer: " Data processing operation has irreducible thermodynamic cost if and only if it is logically irreversible [4]. It seemingly contradicts the von Neumann's principle. Besides, it is not clear how to transmit information with minimal energy.

A clarification was made again by Landauer who considered the case of nonlinear channels in which information is carried in the internal state of a material body (e.g. a bistable molecule with two states separated by a high-energy barrier). He established both for classical [5] and for quantum [6] case that a bit can be transmitted without minimal unavoidable cost. It means that a communication network with only a small loss of transmitted information may require only a small amount of energy for information transmission. However, the problem of organizing an efficient transmission is still to be solved, since the solution proposed by Landauer is a sort of existence theorem and does not suggest a way to design such a transmitter. For example, how to put a carrier of a bit of information into one or another well of a bistable potential? In fact here we face a control problem which is not easy to solve on a molecular level. And despite existence of publications arguing against Landauer's principle (see e.g. [7]) it is clear that there exist physical limits for information transmission and sophisticated methods are needed to approach them.

2. Complexity of future networks does not allow to perform their analysis and design in advance, before their functioning starts. Networks should be created as evolving and adaptive structures that are adjusted during their normal functioning. Although adaptive network concepts are well studied (see, e.g. [8], new challenges demand for fast adaptation which rate is comparable with the rate of information exchange in a network. Analysis and design of such structures are within the scope of the modern control theory. It means that future **network analysis and design paradigms should heavily rely upon control theory and methods.**

A consequence of the above observations is importance of deepening interaction between physics and control theory (in broad sense - cybernetics). However, although both physics and cybernetics were booming in the last century and made revolutionary contributions into science and technology, interaction between two sciences was almost negligible until recently. The reason lies, perhaps, in totally different methodologies of these sciences. Indeed, physics (particularly, mechanics) is a classical *descriptive science* and its aim is to describe the natural behavior of the system. The cybernetics (control science), in the contrary, is a paradigm of *prescriptive science* with the main aim to prescribe the desired behavior of the system [9]. In other words, the purpose of physics is to describe and analyze systems while the purpose of cybernetics is to transform systems by means of controlling action in order to achieve a prescribed behavior.

Surprisingly, the situation has changed in the beginning of the 1990s. New avenue of research in physics was opened by results in control and synchronization of chaos. It was discovered by E.Ott, C.Grebogi and J.Yorke [10] that even small feedback action can dramatically change behavior of a nonlinear system, e.g. turn chaotic motions into periodic ones and vice versa. The idea has become popular in the physics community almost instantaneously. Since 1990 hundreds of papers were published demonstrating ability of small control to change dynamics of the systems significantly, even qualitatively. By the end of the year 2000 the number of quotations of the paper [10] exceeded 1100 while the total number of the papers related to control of chaos exceeded 2500. The number of papers published in peer reviewed journals achieved 300-400 papers per year by the beginning of the XXIst century.

Similar picture can be observed in other fields of interaction between physics and cybernetics: control of quantum systems, control of lasers, control of plasma, beam control, control thermodynamics, etc. Reachability of the control goal by means of arbitrarily small control was observed in a broad class of oscillatory systems [11]

It is important that the obtained results were interpreted as discovering new properties of physical systems. Thousands of papers had been published aimed at studying properties of systems by means of control, identification and other cybernetical means. It is important that overwhelming part of those papers were published in physical journals while their authors represent physical departments of the universities. The above facts provide evidences for existence of the new emerging field of research related both to physics and to control that can be called **Cybernetical Physics** [12, 13]. A concise survey of cybernetical physics is presented in [14].

In this paper the possibilities of analyzing physical systems by means of feedback design are discussed. Firstly, the subject and methodology of cybernetical physics are outlined. Secondly, examples of transformation laws describing the excitability properties of dissipative systems are presented.

2   Subject and Methodology of Cybernetical Physics

Cybernetical Physics (CP) can be defined as the branch of science aimed at studying physical systems by cybernetical means. The problems constituting the subject of CP include control, identification, estimation and others. Due to limited size of the paper only the main class – control problems will be discussed. In order to characterize control problems related to CP one needs to specify classes of controlled plant models, control objectives (goals) and admissible control algorithms. Speaking about the methodology of CP, one needs to classify main methods uses for solving the problems and characterize typical results in the field. A brief description of the subject and methodology of CP is presented below.

2.1   Models of Controlled Systems

A formal statement of any control problem begins with a model of the system to be controlled (plant) and a model of the control objective (goal). Even if the plant model is not given (like in many real world applications) it should be determined in some way. The system models used in cybernetics are similar to traditional models of physics and mechanics with the only difference: the inputs and outputs of the model should be explicitly specified. The following main classes of models are considered in the literature related to control of physical systems. The most common class consists of continuous systems with lumped parameters described in state space by differential equations

$$\dot{x} = F(x, u), \qquad (2.1)$$

where x is n-dimensional vector of the state variables; $\dot{x} = d/dt$ stands for the time derivative of x; u is m-dimensional vector of inputs (control variables). Vector–function $F(x, u)$ is usually assumed continuously differentiable to guarantee existence and uniqueness of the solutions of (2.1) at least at some time interval close to the initial point $t = 0$. It is important to note that model (2.1) encompasses two physically different cases:

A. Coordinate control. The input variables represent some physical variables (forces, torques, intensity of electrical or magnetic fields, etc.) For example a model of a controlled oscillator (pendulum) can be put into the form

$$J\ddot{\varphi} + r\dot{\varphi} + mgl \sin \varphi = u, \qquad (2.2)$$

where $\varphi = \varphi(t)$ is the angle of deflection from vertical; J, m, l, r are physical parameters of the pendulum (inertia moment $J = ml^2/2$, mass, length, friction coefficient); g is gravity acceleration; $u = u(t)$ is the controlling torque. The description (2.2) is transformable into the form (2.1) with the state vector $x = (\varphi, \dot{\varphi})^{\mathsf{T}}$.

B. Parametric control. The input variables represent change of physical parameters of the system, i.e. $u(t) = p - p_0$, where p_0 is the nominal value of the physical parameter p. For example, let the pendulum be controlled by changing its length: $l(t) = l_0 + u(t)$. If $l(t)$ is slowly varying variable, then the model, instead of (2.2) becomes

$$J\ddot{\varphi} + r\dot{\varphi} + m(l_0 + u(t)) \sin \varphi = 0. \qquad (2.3)$$

If the rate of the length change $\dot{l}(t)$ cannot be neglected, it is natural to choose it as the controlling variable:

$$\dot{l}(t) = u(t). \qquad (2.4)$$

In this case the dynamics model derived from Euler-Lagrange equation instead of (2.3) takes the form

$$m(l_0 + u(t))^2 \ddot{\varphi} + 2m(l_0 + u(t))u(t)\varphi + r\dot{\varphi} + mg(l_0 + u(t)) \sin \varphi = 0 \qquad (2.5)$$

and the plant model is described by equations (2.4), (2.5).

Although in some papers the difference between the cases A and B is emphasized, for the purpose of studying the nonlinear system (2.1) the difference is not very important. [1]

If external disturbances are present, we need to consider more general time-varying models

$$\dot{x} = F(x, u, t). \tag{2.6}$$

On the other hand, many nonlinear control problems can be described using more simple affine in control models

$$\dot{x} = f(x) + g(x)u. \tag{2.7}$$

The model should also include the description of measurements, i.e. the l-dimensional vector of output variables (observables) y should be defined, for example

$$y = h(x). \tag{2.8}$$

An important example of output for physical systems is *energy*. E.g. for the pendulum (2.2) the energy is defined as follows: $H = 0.5J(\dot{\varphi})^2 + mgl(1 - \cos\varphi)$. Therefore it is not sufficient to consider only linear functions $h(x)$ as it is accustomed in control theory.

For many systems discrete-time state-space models are used

$$x_{k+1} = F_d(x_k, u_k), \tag{2.9}$$

where $x_k \in \mathbb{R}^n, u_k \in \mathbb{R}^m, y_k \in \mathbb{R}^l$, are state, input and output vectors at kth stage of the process. Then the model will be defined by the mapping F_d. A lot of publications are devoted to control of distributed systems: delay-differential and delay-difference models, spatio-temporal systems described by partial differential equations or their discrete analogs, etc.

2.2   Control Goals

It is natural to classify control problems by their control goals. The conventional control goals are regulation and tracking. State tracking is driving a solution $x(t)$ of (2.1) to the prespecified desired function $x_*(t)$ i.e. fulfillment of the relation

$$\lim_{t\to\infty} [x(t) - x_*(t)] = 0 \tag{2.10}$$

for any solution $x(t)$ of (2.1) with initial conditions $x(0) = x_0 \in \Omega$, where Ω is given set of initial conditions. Similarly, output tracking is driving the output $y(t)$ to the desired output function $y_*(t)$, i.e.

$$\lim_{t\to\infty} [y(t) - y_*(t)] = 0. \tag{2.11}$$

---

[1] It makes sense to treat differently the case of coordinate control and the case of parametric control for linear systems because the linear system with linear parametric feedback control leaves the class of linear systems (becomes bilinear). However the class of nonlinear systems (2.1) is closed with respect to all nonlinear feedbacks.

The problem is to find a control function in the form of *open loop (feedforward) control*

$$u(t) = U(t, x_0), \qquad (2.12)$$

in the form of *state feedback*

$$u(t) = U(x(t)) \qquad (2.13)$$

or in the form of *output feedback*

$$u(t) = U(y(t)) \qquad (2.14)$$

to ensure the goal (2.10) or (2.11).

The key feature of the control problems for physical systems is that the goal should be achieved by means of sufficiently small (ideally, arbitrarily small) control. Solvability of this task is not obvious if the trajectory $x_*(t)$ is unstable, like for the case of chaotic systems, see [10].

A special case of the above problems is stabilization of the unstable equilibrium x_{*0} of system (2.1) with $u = 0$, i.e. stabilization of x_{*0}, satisfying $F(x_{*0}, 0) = 0$. Again, it looks like a standard regulation problem with an additional restriction that we seek for "small control" solutions. However, such a restriction makes the problem far from standard: even for a simple pendulum, nonlocal solutions of the stabilization problem with small control were obtained only recently, see [19]. The class of admissible control laws can be extended by introducing dynamic feedback described by differential or time-delayed models. Similar formulations hold for discrete and time-delayed systems.

Second class of control goals corresponds to the problems of *excitation* or *generation* of oscillations. The goal trajectory $x_*(t)$ is not necessarily periodic. Moreover, the goal trajectory may be specified only partially. In these cases a scalar goal function $G(x)$ is given and the goal is to achieve the limit equality

$$\lim_{t \to \infty} G(x(t)) = G_* \qquad (2.15)$$

or inequality

$$\varliminf_{t \to \infty} G(x(t)) \geq G_*. \qquad (2.16)$$

In many cases the total energy of mechanical or electrical oscillations can serve as $G(x)$.

Third important class of control goals corresponds to *synchronization* (more accurately, *controlled synchronization* as distinct from *autosynchronization* or *self-synchronization*. Generally speaking, synchronization is understood as concordance or concurrent change of the states of two or more systems or, perhaps, concurrent change of some quantities related to the systems, e.g. equalizing of oscillation frequencies. If the required relation is established only asymptotically, one speaks about *asymptotic synchronization*. If synchronization does not exist in the system without control (for $u = 0$) we may pose the problem as finding the control function which ensures synchronization in the closed-loop system, i.e. synchronization may be a control goal. For example the goal corresponding to asymptotic synchronization of the two system states x_1 and x_2 can be expressed as follows:

$$\lim_{t \to \infty} [x_1(t) - x_2(t)] = 0. \tag{2.17}$$

In the extended state space $x = \{x_1, x_2\}$ of the overall system, relation (2.17) implies convergence of the solution $x(t)$ to the diagonal set $\{x : x_1 = x_2\}$. Asymptotic identity of the values of some quantity $G(x)$ for two systems can be formulated as

$$\lim_{t \to \infty} [G(x_1(t)) - G(x_2(t))] = 0. \tag{2.18}$$

Often it is convenient to rewrite the goals (2.10), (2.11), (2.15), (2.17) or (2.18) in terms of appropriate goal function $Q(x,t)$ as follows:

$$\lim_{t \to \infty} Q(x(t), t) = 0. \tag{2.19}$$

For example to reduce goal (2.17) to the form (2.19) one may choose $Q(x) = |x_1 - x_2|^2$. Instead of Euclidean norm other quadratic functions can also be used. E.g. for the case of the goal (2.10) the goal function $Q(x,t) = [x - x_*(t)]^T \Gamma [x - x_*(t)]$, where Γ is positive definite symmetric matrix can be used. The freedom of choice of the goal function can be utilized for design purposes.

Finally, the goals may be specified as modification of some quantitative requirements to the limit behavior of the system, i.e. changing fractal dimension of its attractor.

Some of the above mentioned goals are not typical for conventional control theory because they do not specify the desired behavior of the system completely. These classes of control problems belong to the area of the so called *partial control* which development has become active recently [15, 16]. It is important that the above goals should be achieved without significant intervening the system dynamics, i.e. the control algorithms to be designed should meet the *small control* or *weak control* requirement. Therefore typical formulations of the control problems look as follows:

- find all the behaviors that can be attained by the control functions of the given (sufficiently small) norm;
- find on control function (feedback operator) of minimum norm ensuring the given control goal.

Of course, traditional formulations are not ruled out and may appear in some physical problems.

2.3   Methodology

The methodology of cybernetical physics is based on the control theory. Typically, some parameters of physical systems are unknown and some variables are not available for measurement. From the control viewpoint it means that control design should be performed under significant uncertainty, i.e. methods of robust or adaptive control should be used. A variety of design methods have been developed both for linear and for nonlinear systems [16, 17, 18]. Methods of partial control and weak control are also available [11, 15].

Speed-Gradient Method. Let us describe a fairly general approach to control algorithms design for nonlinear systems: the so called *speed-gradient method*. It is intended for control of continuous-time systems with the control goal specified by means of a goal function. Consider a nonlinear time-varying system and control goal (2.19), where $Q(x, t) \geq 0$ — is a smooth goal function.

In order to design control algorithm the scalar function $\dot{Q} = \omega(x, u, t)$ is calculated that is the speed (rate) of changing $Q_t = Q(x(t), t)$ along trajectories of (2.6): $\omega(x, u, t) = \partial Q(x, t)/\partial t + [\nabla_x Q(x, t)]^T F(x, u, t)$. Then evaluate the gradient of $\omega(x, u, t)$ with respect to input variables $\nabla_u \omega(x, u, t) = (\partial \omega/\partial u)^T = (\partial F/\partial u)^T \nabla_x Q(x, t)$. Finally, the algorithm of changing $u(t)$ is determined according to the differential equation

$$\frac{du}{dt} = -\Gamma \nabla_u \omega(x, u, t), \tag{2.20}$$

where $\Gamma = \Gamma^T > 0$ is a positive definite gain matrix, e.g. $\Gamma = \text{diag}\{\gamma_1, \ldots, \gamma_m\}$, $\gamma_i > 0$. The algorithm (2.20) is called *speed-gradient (SG) algorithm*, since it suggests to change $u(t)$ proportionaly to the gradient of the speed of changing Q_t.

The origin of the algorithm (2.20) can be explained as follows. In order to achieve the control goal (2.19) it is desirable to change $u(t)$ in the direction where $Q(x(t), t)$ decrease. However it may be a problem since $Q(x(t), t)$ does not depend on $u(t)$ directly. Instead one may try to decrease $\dot{Q}$, in order to achieve the inequality $\dot{Q} < 0$, which implies decrease of $Q(x(t), t)$. The speed $\dot{Q} = \omega(x, u, t)$ generically depends on u explicitly which allows to write down (2.20). The speed-gradient algorithm can be also interpreted as a continuous-time counterpart of the gradient algorithm, since for small sampling step size the direction of the gradient is close to the direction of the speed-gradient.

For special case of the system linear in inputs the algorithm (2.20) is nothing but a classical integral control law.

Similarly the following SG-algorithm in finite form is introduced which is a generalization of a proportional control law:

$$u(t) = u_0 - \Gamma \nabla_u \omega(x(t), u(t), t), \tag{2.21}$$

where u_0 is some initial value of control variable, e.g. $u_0 = 0$). More general form of the SG-algorithms can also be used:

$$u(t) = u_0 - \gamma \psi(x(t), u(t), t), \tag{2.22}$$

where $\gamma > 0$ is the scalar gain parameter and vector-function $\psi(x, u, t)$ satisfies the so called *pseudogradient condition*

$$\psi(x, u, t)^T \nabla_u \omega(x, u, t) \geq 0. \tag{2.23}$$

Special cases of (2.22) are *sign-like* and *relay-like* algorithms

$$u(t) = u_0 - \gamma \, \text{sign} \, \nabla_u \omega(x(t), u(t), t), \tag{2.24}$$

where (sign) of a vector is understood component-wise: for a vector $x = \mathrm{col}\,(x_1, \ldots, x_m)$ is defined as $\mathrm{sign}\,x = \mathrm{col}\,(\mathrm{sign}\,x_1, \ldots, \mathrm{sign}\,x_m)$.

In order to make a reasonable choice of the control algorithm parameters the applicability conditions should be developed and checked. A number of applicability conditions can be found in [11, 16]. The main conditions are the following ones: convexity of the function $\omega(x, u, t)$ in u and existence of "ideal" control u_* such that $\omega(x, u_*, t) \leq 0$ for all x (attainability condition).

The speed-gradient algorithm is tightly associated to the concept of Lyapunov function $V(x)$ — a function of the system state nonincreasing along its trajectories. Lyapunov function is an abstraction for such physical characteristics like energy and entropy. I is important that Lyapunov function can be used not only for analysis but also for system design. In particular, the speed-gradient algorithms in the finite form have Lyapunov function the goal function itself: $V(x) = Q(x)$, while differential form of SG-algorithms corresponds to the Lyapunov function of the form: $V(x, u) = Q(x) + 0.5(u - u_*)^{\mathsf{T}}\varGamma^{-1}(u - u_*)$, where u_* – is the desired "ideal" value of controlling variables.

2.4   Results: Laws of Cybernetical Physics

A great deal of the results in many areas of physics are presented in form of *conservation laws*, stating that some quantities do not change during evolution of the system. However, the formulations in CP are different. The results in CP establish how the evolution of the system can be changed by control. Therefore the results in CP should be formulated as *transformation laws*, specifying the classes of changes in the evolution of the system attainable by control function from the given class, i.e. specifying the limits of control. Typical example of transformation law: "Any controlled chaotic trajectory can be transformed into a periodic one by means of control" (OGY law [10]). Another example related to excitation of a system with small control will be presented below.

The term "controllable" in the above context means principal solvability of the problem. To apply the law one needs to use some sufficient conditions ensuring controllability which, however are a matter of further mathematical investigation.

3   Examples of Transformation Laws. Excitability

Many complex systems, including complex networks possess some characteristics that are invariant or decreasing along the trajectories of the nominal (unforced, undisturbed) system. Such functions are analogs of energy for physical (e.g. mechanical) systems and play a important role in system analysis and design. For example, it may be a total utility function in the problems of internet congestion control [23] or energy-like function evaluating the failure threshold [22]. Sometimes a network can be endowed with a generalized Hamiltonian structure which significantly reduces complexity of its analysis [22]. The goal of control in such cases may be to stabilize or to increase the value of the above characteristics and important question is the one about the limitations of such a control. Below the limits of energy control for Hamiltonian and dissipative systems are established which provide an example of transformation laws in nonlinear systems.

3.1   Measuring Excitability of Systems

In [12, 13] the idea to create maximum excitation (resonance) mode in a nonlinear system by changing the frequency of external action as a function of oscillation amplitude was suggested. The corresponding phenomenon in system behavior was termed "feedback resonance". Such a sort of control may be used in a broader class of problems when the goal is to maximize or to minimize the limit value of some energy-like characteristic of a complex system under bounded power of control action.

Consider a system described by state-space equations

$$\dot{x} = F(x, u), y = h(x), \tag{3.25}$$

where $x \in R^n$ is state vector, u, y are scalar input and output, respectively.

To realize the idea of feedback resonance, $u(t)$ should depend on the state of the system $x(t)$ or on the current measurements $y(t)$, which exactly means introducing a state feedback $u(t) = U(x(t))$ or output feedback $u(t) = U(x(t))$. Now the problem is: how to find the feedback law in order to achieve the maximum limit amplitude of output?

Let us pose the problem as that of optimal control: to find

$$Q(\gamma) = \limsup_{\substack{|u(s)| \leq \gamma, \\ 0 \leq s \leq t, \\ x(0)=0, \\ t \geq 0}} |y(t)|^2. \tag{3.26}$$

Assume that the system (3.25) is BIBO stable (bounded input produce bounded output) and $x = 0$ is equilibrium of the unforced system ($F(0,0) = 0, h(0) = 0$) in order to ensure $Q(\gamma)$ to be well defined. Apparently, the signal providing maximum excitation should depend not only on time but also on system state, i.e. input signal should have a feedback form. Note that for linear systems the value of the problem (3.26) depends quadratically on γ. Therefore it is naturally to introduce the *excitability index* (EI) for the system (3.25) as follows:

$$E(\gamma) = \frac{1}{\gamma}\sqrt{Q(\gamma)}, \tag{3.27}$$

where $Q(\gamma)$ is the optimum value of the problem (3.26). It is clear that for linear asymptotically stable systems $E(\gamma) = \text{const}$. For nonlinear systems $E(\gamma)$ is a function of γ that characterizes excitability properties of the nonlinear system. It was introduced in [14] with respect to the energy-like output. For MIMO systems excitability indices E_{ij} can be introduced in a similar way for every pair of input u_i and output y_j. The concept of EI is related to the concept of *input–output (I-O) gain* If I-O gain exists, it provides an upper bound for EI. Conversely, if EI is finite, it estimates the minimal value of I-O gain.

The solution to the problem (3.26) for nonlinear systems is quite complicated in most cases. However we can use approximate locally optimal or speed-gradient solution

$$u(x) = \gamma \, \mathrm{sign} \left(g(x)^{\mathsf{T}} \nabla h(x) h(x) \right), \tag{3.28}$$

where $g(x) = \left. \frac{\partial F(x,u)}{\partial u} \right|_{u=0}$, obtained by maximizing the principal part of instant growth rate of $|y(t)|^2$. It follows from the results of [20] that for small γ the value of $|y(t)|$ achievable with input (3.28) for sufficiently large $t \geq 0$ differs from the optimal value $Q(\gamma)$ by the amount of order γ^2. An important consequence is that excitability index can be estimated directly by applying input (3.28) to the system. For real world systems it can be done experimentally. Otherwise, if a system model is available, computer simulations can be used.

3.2   Properties of Excitability Index

Since excitability index of a system characterizes its sensitivity to a feedback excitation, it is important to relate excitability to inherent dynamical properties of a system. Such kind of bounds for a class of strictly passive systems (systems with full dissipation) were established in [14]. We present here a slightly modified result.

Recall that the system (3.25) is called *strictly passive with dissipation rate* $\rho(x) \geq 0$ if there exists continuous nonnegative function $V(x)$ (storage function) such that for all $t \geq 0$ and any solution $x(t)$ of the system (3.25) the following identity holds

$$V(x(t)) = V(x(0)) + \int_0^t (w(s)^{\mathsf{T}} u(s) - \varrho(x(s))) ds, \tag{3.29}$$

where $w = W(x)$ is an auxiliary output function of the system, the so called *passivity output*.

The storage function $V(x)$ is an analog of energy for the systems of general form (3.25), i.e. identity (3.29) can be interpreted as the generalized energy balance.

Definition 1. Let the set of admissible control consist of functions $u(t)$, bounded for $0 \leq t < \infty$ such that the corresponding trajectories $x(t)$ are bounded. Define *upper and lower excitability indices* of (3.25) *with respect to the output* $V(x)$ as functions $\chi_V^+(\gamma)$, $\chi_V^-(\gamma)$, defined for $0 \leq \gamma < \infty$ as follows:

$$\chi_V^+(\gamma) = \varlimsup_{t \to \infty} \sup_{\substack{|u(\cdot)| \leq \gamma \\ x(0) = 0}} V(x(t)), \tag{3.30}$$

$$\chi_V^-(\gamma) = \varliminf_{t \to \infty} \sup_{\substack{|u(\cdot)| \leq \gamma \\ x(0) = 0}} V(x(t)). \tag{3.31}$$

$\square$

Similarly excitability indices $\chi_y^+(\gamma)$, $\chi_y^-(\gamma)$ with respect to any output $y = h(x)$ are defined. In the case when the input is vector, $u = \mathrm{col}\{u_1, \ldots, u_m\}$

output is also vector $\gamma = \{\gamma_1, \ldots, \gamma_m\}$ and excitability indices are defined as multi-indices. In general case of the system with m inputs and l outputs the excitability indices $\chi_y^+(\gamma)$, $\chi_y^-(\gamma)$ are $l \times m$-matrices depending on m arguments. Note that the technical assumption of boundedness of $x(t)$ can be weakened but we will not do it here for simplicity.

The main result of this section is the following statement.

Theorem 1. *Let system (3.25) be strictly passive and the storage function $V(x)$ and dissipation rate $\varrho(x)$ satisfy inequalities*

$$\alpha_0|w|^2 \le V(x) \le \alpha_1|w|^2 + d, \tag{3.32}$$

$$\varrho_0|w|^2 \le \varrho(x) \le \varrho_1|w|^2 \tag{3.33}$$

for some positive $\alpha_0, \alpha_1, \varrho_0, \varrho_1, d$. Let the set

$$\Omega^- = \left\{ x : W(x) = 0, \ V(x) < \alpha_0 \left(\frac{\gamma}{\varrho_1} \right)^2 \right\}$$

not contain whole trajectories of the free system $\dot{x} = F(x, 0)$.

Then excitability indices $\chi_V^+(\gamma)$, $\chi_V^-(\gamma)$ with respect to $V(x)$ satisfy inequalities

$$\alpha_0 \left(\frac{\gamma}{\varrho_1} \right)^2 \le \chi_V^-(\gamma) \le \chi_V^+(\gamma) \le m\alpha_1 \left(\frac{\gamma}{\varrho_0} \right)^2 + d, \tag{3.34}$$

Besides, the lower bound is realized for the speed-gradient control

$$u(t) = \gamma \operatorname{sign} w(t). \tag{3.35}$$

We see that the action (3.28) or (3.35) creates a sort of resonance mode in a nonlinear system: for weakly damped systems even a small action having form (3.28) leads to large oscillations of the output and can insert a substantial amount of energy into the system. Relations (3.34) can be interpreted as transformation of energy laws for passive systems:

In a strictly passive system with small dissipation of order ρ an energy level achievable by means of control of the level not exceeding γ is of order $(\gamma/\rho)^2$.

3.3   Case of Hamiltonian Systems. Escape from Potential Wells

The above limits for excitability of physical systems by means of controlling actions are expressed in terms of the ratio "(excitation amplitude)/(dissipation)". They can be applied to systems described by Hamiltonian models with possible dissipation:

$$\dot{q}_i = \frac{\partial H(q, p, u)}{\partial p_i}, \quad \dot{p}_i = -\frac{\partial H(q, p, u)}{\partial q_i} - R_i(q, p), \quad i = 1, \ldots, n, \tag{3.36}$$

where $q = \mathrm{col}(q_1, \ldots, q_n)$, $p = \mathrm{col}(p_1, \ldots, p_n)$ are vectors of generalized coordinates and momenta which form the system state vector $x = \mathrm{col}(q, p)$; $H = H(q, p, u)$ is the Hamiltonian function for controlled system; $u(t) \in \mathbb{R}^m$ is controlling input function; $R(q, p) = \mathrm{col}(R_1(q, p), \ldots, R_n(q, p))$ is the dissipation function, satisfying the inequality

$$R(q, p)^\top \frac{\partial H_0(q, p)}{\partial p} \geq 0, \tag{3.37}$$

where $H_0(q, p) = H(q, p, 0)$ is the energy of free system. In what follows we assume that the Hamiltonian is linear in control: $H(p, q, u) = H_0(p, q) + H_1(p, q)^\top u$, where $H_0(p, q)$ is the internal Hamiltonian and $H_1(p, q)$ is an m-dimensional vector of interaction Hamiltonians.

Hamiltonian description is typical for many physical systems, including molecular systems. There are also attempts to apply Hamiltonian models for communication networks [22]. The inequality (3.37) means dissipation of the energy over trajectories of the free (uncontrolled) system: $\dot{H}_0 \leq 0$. It means that systems (3.36) are passive in the sense of (3.29) and Hamiltonian of the free system can be chosen as a storage function: $V(x) = H_0(q, p)$, while passivity output is the Poisson bracket $W(x) = [H_0, H_1]$ of smooth functions $H_0(p, q)$ and $H_1(p, q)$ defined in a standard manner:

$$[H_0, H_1] = \sum_{i=1}^{n} \left(\frac{\partial H_0}{\partial p_i} \frac{\partial H_1}{\partial q_i} - \frac{\partial H_0}{\partial q_i} \frac{\partial H_1}{\partial p_i} \right).$$

As an example consider the problem of escape from a potential well which is important in many fields of physics and mechanics. Sometimes escape is an undesirable event and it is important to find conditions preventing it (e.g. buckling of the shells, capsize of the ships, etc.). In other cases escape is useful and the conditions guaranteeing it are needed. Escape may correspond to a phase transition in the system. In the area of information physics briefly described in the Introduction, escape may correspond to transition from the state "0" to state "1" of the information system, i.e. to creation of a bit of information. In all cases the conditions of achieving escape by means of as small external force as possible are of interest.

Consider nonlinear oscillators with one degree of freedom, modeled as

$$\ddot{\varphi} + \varrho\dot{\varphi} + \Pi(\varphi)' = u, \tag{3.38}$$

where $\varrho > 0$ is the damping coefficient. Equation (3.38) can be transformed to the Hamiltonian form with coordinate and momentum $q = \varphi$, $p = \dot{\varphi}$, the Hamiltonian function (energy) $H_0(\varphi, \dot{\varphi}) = \frac{1}{2}\dot{\varphi}^2 + \Pi(\varphi)$ and passivity output p.

In [21] such a possibility (optimal escape) has been studied for typical nonlinear oscillators with a single-well potential $\Pi_e(\varphi) = \varphi^2/2 - \varphi^3/3$ (so called "escape equation") and a twin-well potential $\Pi_D(\varphi) = -\varphi^2/2 + \varphi^4/4$ (Duffing oscillator). The least amplitude of a harmonic external forcing $u(t) = \overline{u} \sin \omega t$ for which no stable steady state motion exists within the well was determined by intensive computer simulations. For example, for escape equation with $\varrho = 0.1$ the optimal amplitude was evaluated as $\overline{u} \approx 0.09$, while for Duffing twin-well equation with $\varrho = 0.25$ the value of amplitude was about $\overline{u} \approx 0.21$.

Using feedback forcing we may expect reducing the escape amplitude. In fact using the locally optimal control

$$u = \gamma \mathrm{sign}(\dot{q}), \tag{3.39}$$

the amplitude of feedback leading to escape can be easily calculated, just substituting the height of potential barrier $\max_{\Omega} \Pi(\varphi) - \min_{\Omega} \Pi(\varphi)$ for $\overline{H}$ into (3.34) where Ω is the well corresponding to the initial state, see [12]. For example, in the case of escape equation $\overline{H} = 1/6$, $\varrho = 0.1$ and $\overline{u} = 0.0577$, while for Duffing oscillator with $\overline{H} = 1/4$, $\varrho = 0.25$ escape amplitude is estimated as $\overline{u} = 0.1767$. The obtained values are substantially smaller than those evaluated in [21]. The less the damping, the bigger the difference between the amplitudes of feedback and nonfeedback signals leading to escape. Simulation exhibits still stronger difference: escape for Duffing oscillator occurs for $\overline{u} = \gamma = 0.122$ if the feedback (3.39) is applied. Note that the oscillations in the feedback systems have both variable frequency and variable shape.

In [12] the dependence of escape amplitudes on the damping was also studied by means of computer simulations in the range of damping coefficient ϱ varying from 0.01 to 0.25. Simulations confirmed theoretical conclusion that the feedback input amplitude required for escape is proportional to the damping. We may evaluate the efficiency of feedback μ as the ratio of escape amplitudes for harmonic ($\overline{u}_h$) and feedback ($\overline{u}_f$) forcing: $\mu = \overline{u}_h / \overline{u}_f$. Then the efficiency of feedback is inversely proportional to the damping for small values of damping.

4   Conclusions

Now, after a decade of vigorous development, cybernetical physics is still an emerging field. In the paper the limits for the speed-gradient control of energy-like characteristics have been demonstrated providing an example of transformation law for dissipative systems.

The above mentioned results provide also a good starting point to apply speed-gradient method to control of failures in complex networks based on generalized Hamiltonian model of [22]. Speed-gradient control of escape from potential wells may be helpful to control of molecular objectives, necessary to achieve physical limits of information transmission.

An interest of control community to control problems for communication networks has increased during recent years significantly [23, 24, 25]. Other existing methods of nonlinear control are under further development to become applicable to new important problems.

References

1. Landauer, R. The physical nature of information. Phys. Lett. A. **217** (1996) 188-193.
2. Shannon, C. A Mathematical Theory of Communication. Bell Syst. Tech. J. **27** (1948) 3, 379-423; 4, 623-656.
3. von Neumann, J. Theory of Self-Reproducing Automata, edited and completed by Arthur W. Burks, University Illinois Press, Urbana and London (1966).
4. Landauer, R. Irreversibility and heat generation in the computing process. IBM J. Res. Develop **3** (1961) 183191.
5. Landauer, R. Energy Requirements in Communication. Appl. Phys. Lett. **51** (1987) 2056-2058.
6. Landauer, R. Minimal Energy Requirements in Communication, Science **272** (1996) 1914-1918.
7. John D. Norton, Eaters of the Lotus: Landauer's Principle and the Return of Maxwell's Demon", PHIL-SCI 1729 http://philsci-archive.pitt.edu/archive/00001729/
8. Bartsev S.I., Okhonin V.A. Adaptive networks of information processing. Krasnoyarsk, Inst. of Physics SO AN SSSR, 1986 (In Russian).
9. Brockett R.W. Control theory and analytical mechanics. Geometric Control Theory, Lie Groups. V. VII / Eds. C. Martin, R. Hermann. Mat. Sci. Press, Brookine, MA, (1977) 1-48.
10. Ott T., Grebogi C., Yorke G. Controlling chaos. Phys. Rev. Lett. **64** (1990) 11 1196-1199.
11. Fradkov A.L., Pogromsky A.Yu. Introduction to control of oscillations and chaos. World Scientific, Singapore (1998).
12. Fradkov A.L. Exploring nonlinearity by feedback. Physica D. **128** (1999) No 2-4 159-168.
13. Fradkov A.L. Investigation of physical systems by means of feedback. Autom. Remote Control **60** (1999) 3 3-22.
14. Fradkov A.L. Cybernetical physics. Nauka, St.Petersburg 2003, 208p.(In Russian).
15. Vorotnikov V.I. Partial Stability and Control Birkhäuser (1998).
16. Fradkov A.l., Miroshnik I.V., Nikiforov V.O. Nonlinear and adaptive control of complex systems. Dordrecht: Kluwer Academic Publishers (1999).
17. Krstić, M., I. Kanellakopoulos and P.V. Kokotović. Nonlinear and Adaptive Control Design, New York, Wiley (1995).
18. Zhou K., J.C.Doyle and K.Glover. Robust and Optimal Control Prentice Hall (1996).
19. Shiriaev A.S. and A.L.Fradkov. Stabilization of invariant sets for nonlinear systems with application to control of oscillations. Intern. J. of Robust and Nonlinear Control, **11** (2001) 215-240.
20. Chernousko F.L. Some problems of optimal control with a small parameter. J. Appl. Math. Mech. (1968) 3 12-22.
21. Stewart, H.B., Thompson, J.M.T., Ueda, U. & Lansbury, A.N. Optimal escape from potential wells – patterns of regular and chaotic bifurcations. Physica D **85** (1995) 259-295.

22. DeMarco C.L. A phase transition model for cascading network failure. IEEE Control Systems Magazine, **21** (2001) 6 40-51.
23. Law S., Paganini F., Doyle J. Internet congestion control. IEEE Control Systems Magazine, **22** (2002) 1 28-43.
24. Wen, J.T., Arcak M. A unifying passivity framework for network flow control. IEEE Trans. Autom. Contr. **AC-49** (2004) 2 162-174.
25. Special Issue on Networked Control Systems. Eds. P.Antsaklis and J.Baillieul. IEEE Trans. Autom. Contr. **AC-49** (2004) 9.

Panel Report: "Main Principles to Guide R&D in Algorithms, Protocols and Middleware"

Ioannis Stavrakakis[1] and Fabrizio Sestini[2]

[1] University of Athens
ioannis@di.uoa.gr
[2] EU Commission
Fabrizio.Sestini@cec.eu.int

Abstract. The panel objective was to discuss with the audience and highlight commonalities and inter-dependencies between the papers presented in the first two sessions of WAC 2004 - **Network Management** and **Models and Protocols**. As this panel was the first one, main discussion tried to define autonomicity, address the more general questions around this new concept and distinguish it from previously introduced and studied concepts.

1 What Is Autonomicity?

Do we need a new formalism, unifying abstraction or new paradigm to describe and manage it?

There was some consensus on the need for a scalable policy language in a highly distributed-based networking scenario, whereas references to agent technology or to active and intelligent networking should be avoided. However fundamental aspects that lead to new networking paradigms need to be effectively addressed first, such as the behavior and cooperation of autonomic entities (based on considerations of economics and game-theoretic approaches) and the self-organizational aspects (and associated effective algorithms).

The issue is not simply the individual autonomic behavior serving self-interest, but how these autonomic, self-driven, individual behaviors lead to a desirable, acceptable global behavior (this is addressed in biology in the context of behavioral ecology). In this respect feedback from the network and control aspects are also important.

We should also clearly articulate to what extent autonomic communications is different from re/auto-configurability, which should be considered as just one aspect of it.

2 Why Do We Need Autonomic Communications?

Internet is indeed complex, but is working pretty well and satisfying a lot of requirements. The following three reasons in favour of development of autonomic communication were expressed during the panel.

M. Smirnov (Ed.): WAC 2004, LNCS 3457, pp. 269–270, 2005.

First, to cope with the increasing complexity of the Internet by enabling autonomicity, i.e. new networking solutions that facilitate injecting an autonomic component into the network and figuring out automatically what to do with it.

Second, to share the cost of managing networks (this comes from a similar reasoning as to why autonomic computing is needed – IBM's desire not to be burdened with the cost of mainframes).

Third, the cell size is decreasing in mobile wireless networks and such networks proliferate. It is therefore highly desirable to deploy and manage them autonomously and not manually, to reduce the increasing management cost.

3   Is Autonomicity More of a Revolutionary Than an Evolutionary Concept?

To claim that this research is more revolutionary as opposed to evolutionary, it was suggested not to concentrate only on the current or emerging problems it may help solve, but rather try to establish if it fits a possible trend, by looking fairly far into the past of this technology and trying to foresee its future.

60 years ago networks had one or very few owners (in a country or a continent) and this worked well under the circuit-switching paradigm (which did not see any revolution over those years). 30 years ago, the revolutionary packet-switched Internet technology was formed and is managed now by a very large number of distributed entities. Today, we have end users being capable of providing network resources (as routers, storage as content proxies) to networks and becoming themselves the network. Thus, it is conceivable that, several years in the future; the network will basically be built, owned and run by the millions of autonomic end-users. Getting to and managing such networks requires revolutionary steps, which can justify why we need autonomic communications. The trend suggests that it is very likely that autonomic nodes will shape the networking environment in the future, contributing their increasingly powerful resources to it.

Consequently, in addition to developing the concept of autonomicity in order to re-design networking to be more effective in the current and emerging networking environment, we should develop this concept in such a revolutionary way as to capture early enough emerging trends that will lead to entirely new networking environments where autonomic communications would clearly be the paradigm to employ.

Panel Report: "Grand Challenges of Network and Service Composition"

Giuseppe Valetto[1] and Fabrice Saffre[2]

[1] Telecom Italia Lab
Giuseppe.Valetto@tilab.com
[2] British Telecom
fabrice.saffre@bt.com

Abstract. This brief report intends to summarize some of the things we learned during WAC2004 sessions "Network Composition" and "Negotiation and Deployment", and that were highlighted during the subsequent panel discussion. We try to focus especially on the following aspects: traits and trends of convergence emerging from the rather diverse findings presented in those sessions; controversial or divergent opinions on some of those findings; open issues that should be addressed by the autonomic communication community and how to tackle them; and major research directions that seem likely to emerge and shape a significant part of the autonomic communication landscape.

1 Findings and Commonalities

We notice how a certain set of issues surfaced repeatedly in both sessions. Fig. 1 attempts to draw some relationships among those issues, starting from the idea of *Adaptation* as a generic term for the self-regulating operation of an autonomic system. Several of the works presented deal with *Composition* as an important form of adaptation and a first-class autonomic primitive. *Policies* and *Negotiation* are widely regarded as useful means to strategize about and achieve adaptation through composition, in a top-down vs. bottom-up fashion: policies can dictate the terms of composition (e.g. *when, what and how*), while negotiation can be used to spontaneously reconcile competing and/or conflicting policies, and converge towards stable system configurations. Finally, *Semantic Knowledge* can be used to obtain an explicit and abstract representation of autonomic adaptation, including composition. That representation remains formal and hence can be distilled into policies, but at the same time provides the ability to understand discuss, communicate and review the autonomic behavior of a system.

Another evident common element is that wireless communication provides major motivation and an obvious test bench for the investigation of autonomic paradigms: LANs, PANs, sensor networks, as well as ubiquitous, ambient and ad hoc networking, are among the wireless and mobile contexts that look like natural catalysts for

M. Smirnov (Ed.): WAC 2004, LNCS 3457, pp. 271–274, 2005.

autonomic capabilities. It seems also evident that all of them require some kind of context-awareness (e.g. location-, user-, service-awareness) embedded in the autonomic communication facilities, possibly as part of the knowledge base.

Other agreed-upon, significant enablers of autonomic communication are transparent addressing, seamless handover, strong decentralization of all adaptation mechanisms, and a regard for issues like resource, security and trust management as first-class elements.

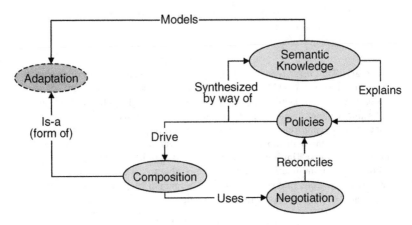

Fig. 1. Common issues and their relationships

2 Divergences

Along many of the presentations and discussions, the issue of *emergent behavior* was often either explicitly mentioned, or alluded to. However, we have recorded diverse opinions on its actual relevance to the problems in autonomic communication. On the one hand, it seems clear that it has a clear appeal, because of its resonance with the biologic metaphor at the very basis of all things autonomic, and its promise to enable complex, numerous and strongly autonomous elements that co-exist in the same environment to act in concordance and with a common sense of purpose. On the other hand, there is a feeling that the biologic metaphor should not be over-stretched, and a concern that such an approach – as well as many evolutionary or "cognitive" approaches - could break once applied to communication infrastructures and services, under their extremely demanding timeliness and predictability requirements.

In a similar fashion, different opinions exist on whether autonomic features should originate from explicit vs. implicit provisions (e.g., dedicated protocols that account for and codify features such as robustness, flexibility and fault-tolerance vs. spontaneous or stigmergetic interactions among communication elements that converge towards a mutually sustainable and satisfactorily functional configuration).

3 Open Issues and Next Steps

A number of suggestions and questions have been raised, trying to indicate how to focus the undergoing work and discussion of the autonomic communication community towards a set of incremental goals that can help map this largely uncharted territory.

A major issue seem to be trying to define the boundaries of the autonomic communication domain: it seems obvious that it comprises self-* issues within a network infrastructure, as well as at the juncture of different networks; it is also evident that it extends to a degree up to the level of the services carried by those networks, but what kind of services are going to be affected the most – and in what application domains - is not equally well understood.

A related issue is the categorization of the techniques that are relevant for autonomic communication: in Session 2A and 2B – and throughout the Workshop in general – a wide spectrum of techniques, ranging from control theory (with its hard mathematical foundations), to bio-inspired techniques (for instance, emergent behavior), and from distributed negotiation algorithms to knowledge-based reasoning have been presented and debated. It must be better understood what solutions are suitable for what problems, in a multi-dimensional space that includes qualitative parameters, such as problem type, attainable scale, level of impact, as well as quantitative properties, such as timeliness, predictability, reliability.

Another open issue of great interest of the community is the level of *transparency* that autonomic communication facilities shall strive for. Transparency is in itself a multi-faceted concept, since it encompasses technical factors (e.g., non-intrusiveness in affected systems) as well as human-observable factors (e.g., the ability to understand, analyze and influence autonomic behavior on the part of technicians, as well as seamless and "hidden" operation from the point of view of end users).

An operational suggestion to address some of the questions above is to work together to propose and formalize a sort of "autonomic communication reference problem" (or problem set), against which proposed approaches should be evaluated, independently of their application domain. The reference problem might be characterized as a "check-list" of observable and demonstrable autonomic features, which an autonomic communication system should strive to address, thus enabling a form of comparison with respect to other solutions.

4 Research Directions

Research issues in autonomic communication are likely to revolve around achieving two complementary objectives: identifying suitable design principles and testing implementation/deployment strategies. As for the first, many sources of inspiration are available from the natural world including, but not limited to, biological systems. As for the second, a number of well-established domains in computer science (machine learning, formal methods etc.) offer reliable tools and a substantial body of knowledge to start with. This is not the place to review these, however, so we chose to sketch a picture of what we believe will be a fundamental common denominator to all future research in the field of autonomic systems.

Traditional engineering starts by specifying desirable system-wide characteristics and then designs/selects individual components under the assumption that the whole is only the sum of its parts. In extremely large distributed systems, this "top-down" approach is under severe strain to deliver viable solutions, which is a major rationale for autonomic communication. Yet this new paradigm raises issues of its own, mostly due to the apparition of complexity (in the restrictive scientific sense of emergent global properties within large ensembles of interacting units). However, despite being aware of this difficulty, many technologists seem reluctant to cross the cultural barrier between a proven and immensely successful paradigm (inherited from the industrial revolution) and the newer science of complexity, which is less well understood by engineers.

Complexity science provides powerful methods for dealing with probabilistic predictability and describing in a rigorous and useful way systems comprised of individually unpredictable elements. Over the last three decades, it has been extensively demonstrated that variability in the individual response of its constituents does not necessarily translate into the frequency distribution of a system's states exhibiting a similar amount of 'noise'. On the contrary, the huge number of interactions and the presence of intricate feedback loops often mean that the system as a whole can only exist in a limited number of configurations, despite the largely random behavior of individual units. The science of complexity mainly consists of identifying these configurations, determining their probability of occurrence, and understanding/characterizing transitions between them and trajectories leading to them (e.g. bifurcation).

The sheer size of a large network comprised of many thousands of components means that the state of a large distributed computing environment will virtually always be the result of an unforeseeable combination of many events, and so can be described best probabilistically. While there may be an increased recognition of this situation, there is a poor awareness of the methods capable of dealing with it. The heterogeneity of the underlying infrastructure (in terms of purpose, capability, and ownership) precludes a centrally imposed set of rules defining the function and privileges of every participant. Instead, we must find ways to engineer autonomic principles, like self-configuration, into individual elements and their interactions, so as to allow them to deal with unexpected situations, requests, combinations of events, etc.

Complex systems theory and modelling can and must help us understand which macroscopic behaviour is more or less likely to emerge from the many interactions between heterogeneous devices. The real challenge is not to cope with microscopic unpredictability - the conceptual tools required to handle its macroscopic effects are readily available. The difficulty resides in identifying and weighting the factors involved, so that the purpose of fine-tuning the local rules is not defeated by the presence of 'hidden variables' capable of pushing the entire system into an unexpected/undesirable state.

Panel Report: "How the Autonomic Network Interacts with the Knowledge Plane?"

David Lewis

Knowledge and Data Engineering Group,
Trinity College Dublin,
Dublin 2, Ireland
Dave.Lewis@cs.tcd.ie

Abstract. This panel was held at the end of the Workshop on Autonomic Communication Principles, on the 19th October 2004. It brought together speakers from session 3 on Resilience and Immunity and session 4 on Meaning, Context and Situated Behaviour. The panellist were Anuarg Garg (University of Trento), Fabio Massacci (University of Trento), Christian Tschudin (University of Basel), Simon Dobson (University College Dublin), Maurice Mulvenna (University of Ulster) and Cesar Santivanez (BBN Technology).

1 Panel Report

The panel opened with a question from the audience asking how the evolvability of Autonomic Communications (AC) and the stability of the resulting architectures and systems can be ensured. The panel responded stating that stability can not be regarded in terms of deterministic system configuration, but needs to be viewed in terms of behavioural stability. Thus we must tolerate a level of volatility but only within a well understood behavioural envelope that relates to specific autonomic tasks. In other words, we should focus on enforcing specific bounds on the adaptivity that self-managed systems may exhibit, rather than on achieving full behavioural determinism. With respect to the evolution toward and evolvability of AC, it was agreed that gradual changes were a real-life necessity. As a result we require ways to subdivide AC architectures into separate areas of concern that can be attacked, solved and deployed independently. However, there were no immediate suggestions for the lines along which such a separation would best be made. Preceding the panel, a poster presentation had included a synthesis of issues raised during the workshop in the form of a layered cube reminiscent of that used to explain broadband ISDN principles during the 1990's. It was observed that this synthesis served to show the potential complexity and inter-connectiveness of issues in AC. Reactions to this model, however, also hinted at the challenges in defining any clear architectural separations for AC given our current understanding of the field. It also spurred comments on the lessons that could be learnt by the failure of ATM to reach its technical potential due to a lack of flexibly in reacting to changing economic and market concerns. There was

M. Smirnov (Ed.): WAC 2004, LNCS 3457, pp. 275–278, 2005.

broad consensus that these lessons must be heeded by the AC community in considering any evolution strategy.

Next a speaker from the audience observed that the success of the Internet was due in no small part to the clear separation of the application from the network via a simple interface, but that this separation also potentially limited the evolution of communication services. The question then posed asked whether the application-network separation should be subject to some re-integration to open the door to fundamental reappraisal of architectures suitable for autonomic communication. Such re-integration is already a strong feature in much current research in cross layer optimisation for wireless and ad hoc networks. Some panellists viewed that the application/network separation should not be violated due to its significance in allowing application innovation. Another emphasised the need for some form of modularity in order to allow the problem to be broken down and for innovation and competition to be encouraged. A further response questioned the assumption that the knowledge needed for AC should not be allocated to a separate 'plane' and suggested that instead it should be integrated with the data plane of the network. It seemed increasingly apparent from these responses that a layered architecture with well defined interfaces between layers was not readily apparent for AC. Instead, it was observed that the focus should shift to the adaptive sharing of information across conventional network boundaries, but in a way that was constrained by business, regulatory, or task concerns, rather than the need to have a fixed interface in support of a stratified architecture. However, as a result, the computational elements that populate such a loosely structured AC architecture must be more able to deal with information exchanged with other elements without pre-programmed understanding of its semantics.

The next speaker from the floor reinforced this view by observing the use of terms such as 'network of workflows' and 'architecture as a program' in the workshop. This was followed by a specific question on how AC systems can best determine when 'text becomes context', i.e., how is available knowledge to be judged useful context for a problem? Here the panel was broadly agreed that there is no canonical model of what represents context for AC tasks. Instead, context had to be formed on a subjective basis by AC elements, resolving their knowledge needs against the information that is available and accessible to them. This raises the prospect that the process that identifies and uses information as context determination is itself context-aware.

The final speaker from the floor asked how AC systems could be made conceptually simple. There was consensus from the panel that the problem domain was implicitly complex, and that the target should be to simplify the human experience of the management of complex communication services. It was observed however that we should not aim for one-size-fits-all approach to exposing complexity to the human administrator, but to instead aim for complexity on demand to reflect the tasks, skills, and cognitive abilities of the individuals concerned. However, though the complexity that individual AC components expose could be minimised, this is likely to be at the expense of increasing complexity in how such components interact.

The panel ended with each panellist expressing how what they had learnt in the workshop would impact on how they might subsequently present their papers. Christian Tschudin, whose paper presented a fine-grained approach to integrating code fragments that resulted in robustness and self-healing properties of the overall program, had his views on the need for a bottom-up approach reinforced. Fabio Massacci, who had presented a paper on negotiating the knowledge exchange needed to resolve access control policies, was interested in applying such a fine-grained approach to achieving robustness in policy integration. Simon Dobson, whose paper examined the role of contextual semantics in AC, would address the role of composition more carefully in the determination of the semantics of context. Maurice Mulvenna, whose paper had addressed the customisation context knowledge to the task at hand, was interested in the need for a more rigorous experimental approach to AC development. Anuarg Garg, whose paper had addressed a peer-to-peer trust mechanism combining concepts of reputation, quality, and credibility, expressed a need to more clearly define the relationship between P2P and AC. Cesar Santivanez, whose paper addressed adaptable ad hoc networks, saw the need to make ad hoc networks more application aware.

2 Conclusions

In conclusion, the role of a 'knowledge plane' in relation to Autonomic Communication remains unclear, in no small part due to a lack of consensus on what characterises such a plane. Clark et al's 2003 SIGCOM [1] paper described a 'Knowledge Plane for the Internet' as operating in parallel to existing concepts of data, control, and management planes. However, a closer examination of this work reveals that it encompasses not only knowledge monitoring and analysis but also its use for the planning and execution of network control and management tasks, thus making it much closer in functional scope to Autonomic Communications. Their use of the 'knowledge plane' metaphor probably owes more to the pragmatic tendency in the Internet Community to progress through a set of small, individually motivated steps rather than as part of a larger cohesive vision. Though the panel recognised the need for incremental evolution of AC, the aim of the workshop was to start work on a comprehensive AC vision and on the research agenda needed to realise it. As such, we are justified in questioning the core separation of layers and planes underlying the design of current networks, and in particular the persistence of this mindset into architecture for AC. The panel underlined this critical stance, raising the prospect that the AC domain may not be amenable to decomposition into the type of orthogonal separations that has guided the separation of concerns in current networks. This has profound implications for the AC research agenda and the resulting market in AC systems. Though an alternative architectural structure is not yet evident, some themes have been hinted at in the panel. These include the need: for composition of AC elements; for mechanisms to bound the adaptive behaviour of such compositions, and for mapping this adaptive behaviour to bounds on the behaviour of elements. Also raised is the need to tailor both the exposure of complexity and the employment of contextual knowledge, to the specific task at hand.

References

1. Clark, D., Partridge. C., Ramming, J.C., Wroclawski, J.T. "A Knowledge Plane for the Internet", in Proc. of SIGCOMM'03, 25-29 August 2003, Karlsruhe, Germany

Author Index

Lecture Notes in Computer Science

For information about Vols. 1–3477

please contact your bookseller or Springer